The Pocket Bo
PLUMBIN(

CW01335861

The Pocket Book of
PLUMBING

Jan Woudhuysen

Evans Brothers Limited London

Published by
Evans Brothers Limited
Montague House
Russell Square, London, WC1B 5BX

Text © Jan Woudhuysen 1980
Illustrations © Evans Brothers Limited 1980

First published 1980

Drawings by Illustrated Arts

British Library Cataloguing in Publication Data

Woudhuysen, Jan
  The pocket book of plumbing.
  1. Plumbing – Amateurs' manuals
  1. Title
  696'.1   TH6124

  ISBN 0-237-44999-4

Printed in Great Britain by
T. & A. Constable Ltd, Edinburgh
PRA 6833

Contents

Introduction

This book differs from almost every other do-it-yourself book on plumbing in that it is intended for the absolute beginner. I've assumed that you really know nothing about plumbing, about how to use the special plumbing tools, how to undo nuts, or even how to tell the difference between a nut and a bolt.

After all, it's no good for the beginner to be told: 'Step 2: Undo the shield of the tap, and loosen the gland nut before packing the gland'. Unless you know quite a bit about plumbing, the terms *shield*, *gland nut* and *packing* are meaningless, nor is it clear which way you are to turn the gland nut, and with what tool.

Yet if someone were willing to sit down and explain, just once, exactly what is involved, it would be possible to tackle tap-washer-changing as well as larger projects.

If at times you find this book a little too simple, please bear with me patiently. There are very many readers who will not find it too simple, but will be pleased to learn at long last exactly which way round one pushes a spanner to undo a nut. Many tasks which 'everybody knows how to do' are in fact done badly, every time, even by professionals. So follow the instructions carefully, and read all of them, if only once through.

Actually, plumbing is much less difficult than you might imagine if you watch a plumber repairing an old sink. Plumbing used to be a very skilled job which required lots of training and practice in order to make perfect joints in old-fashioned lead pipes. Now we use copper tubes, pre-formed brass or copper joints, plastic wastes with push-on fittings, and all of these can easily be handled by the amateur. It used to be true that a plumber needed a lot of strength in order to heave cast-iron baths, heavy clay-ware sinks and cast-iron pipes up and down the stairs; now we use plastic baths and stainless-steel sinks. Specialist plumbing skills are still needed to deal with more complicated tasks; generally speaking, the reader needn't worry, as those jobs won't be dealt with in this book.

Each job which is included attempts to be complete. I shall list the materials and tools you'll need, the time it will probably take, and a detailed account of the procedure.

Try to have some fun while you're at it. Children can watch taps to see if things go off, or if they leak. They can hand you tools and pull up your sleeves when you want to wash your very dirty hands. Your friends can read you out the next bit from this book, or they can nip round for the (forgotten) extra fitting from the plumber down the road. If you have a big job going on, arrange to have supper and a bath with friends at their house.

1 Advantages and disadvantages of doing it yourself

I think I can best gain your trust by being honest; that way you'll believe me when I say that you can do a job yourself.

I also think it is a good idea, before you start any piece of work, to see how much it will *really* cost you, both in money and in time. If you then get a quote from a plumber, you may find that it is hardly worth the bother of doing it yourself; alternatively, you may find that it saves you so much money that it will keep you cheered during all the weeks of work and frustration.

A plumber's price will consist of labour costs, materials costs, and overheads and profits. Each of these factors is about one-third of his total bill. Do not assume, though, that by doing it yourself you will save two-thirds. The contractor buys his materials at a considerable discount, and you will have to buy or hire tools. It *is* safe to assume, however, that on average you will save about 50%–55%, and even that sounds quite good.

Another, less obvious, advantage is that by placing pipes exactly where you want them, by burying pipes in the thickness of the plaster on the wall, by cutting neat holes in the kitchen cabinets exactly where they are needed, you may well finish with a much neater job than a plumber might have done. Some plumbers do not seem to care about these matters, and will cheerfully hack holes in expensive kitchen units with a hammer and screwdriver – as if key hole saws had never been invented. You can also, by doing your own tiling, electrical work and carpentry as well as plumbing, integrate the trades (plumbing, carpentry, tiling, etc.) in such a way as to achieve that really well-fitted bathroom or kitchen which is normally seen only in magazines, with a price tag of 'around £10,000, Madam'.

Then, too, the job you do yourself can be done when *you* want it done, and not just when the busy plumber has a spare moment. Often the month or two you have to wait for him to turn up more than outweighs the speed (his three days to your two months) with which he does the job.

The biggest disadvantage in doing it yourself, though, is that it does take very much longer, especially if you work at it only at evenings and weekends. A job which might take a plumber two days might take you two weeks; being without a bath or w.c. for that period is not so convenient. Careful planning, the purchase of some extra fittings (as a precaution), and the use of a good handbook will help, as will the co-operation and kindness of helpful friends and neighbours.

A further disadvantage is that when you do the job yourself, you may finish up with something you feel a *real* plumber might have done much better. You may console yourself with the thought that many professionals make a very sloppy job. And in any case, if you know in advance that the pipework will look as though it's been done up in haste and repented at leisure, you can always arrange matters in such a way that the pipes are hidden by some knotty-pine panelling or by Polyfilla.

The final disadvantage in plumbing it yourself is that it's a lot of hard work. You'll finish with sore muscles, grazed knuckles, raw elbows. Not always, but often enough. But for all that, plumbing it in yourself can be fun – and you may be very pleasantly surprised in these inflationary times to realise how much money you have saved.

2 Getting it done professionally

There are three situations in which it may be better for you to call on the services of a professional plumber: The job may be beyond your scope, or beyond the scope of a simple book like this one; you may have costed the whole job, and decided that in view of the cost of materials, tools, time taken off work, and the general inconvenience of not having a bath for two months, it is uneconomic to do it yourself; you may have tried doing some of the simpler jobs in this book, only to find that you are the world's worst plumber.

Hiring a plumber, whether to do a simple replacement of a tap or the complete renewal of the cold water system, requires some care. There follows a guide on how to do this wisely:

Select your plumber Try to find one who has given satisfaction on a previous job. Ask to look at the job, so you can check his standard of workmanship. Check his price against that of a plumber who hasn't given satisfaction, so you'll know if it's worth paying extra.

If you don't know anyone in the area, or if nobody has recently used a plumber, try a building site. Choose a small one, perhaps a site where they're building or converting one house or flat. Most small builders know several local plumbers; if they are too expensive, or too sloppy, they don't get used again. Ask the foreman.

Often, if there's a plumber on a site, he may be looking for some spare-time evening work. If the site is near your home, you may get the job done – after hours – quickly and cheaply. If there are no local friends or building sites, a chat with an enterprising estate agent might produce good results.

Whatever you do, don't hang on to those cheaply-printed cards with 'Emergency 24-hour Plumbing' written all over them. The worst emergency may be when you have to pay.

Write down, simply and carefully, all the jobs you want done Number each item, and put it on a separate line, like this:

> 5. A new tap on the downstairs lavatory basin
> 6. Change the w.c. pan on the first floor

Ask the plumber to quote a separate price for each item. If you change your mind, or if you have to leave an item out because the total price is too high, you can choose which one to leave out. This is useful too if you want to add an item which has already been given a price, e.g. you might want to put in two basins in a bathroom, instead of one.

If you add to the list, or take something away, don't be too rigid about keeping to these prices. Any change in your list may cost up to 10% more or less than the price shown, because of complications which your plumber will explain – if you will listen without getting angry. There can be many reasons why this is so, as varied as the houses in which we live.

Agree with the plumber, before he starts, the following:
> For any job which can't be assessed before it's started: a) his hourly rate, and b) the limit to which he can work before further permission from you is required.
> The total cost of priced items.
> When he will start, and how soon he will finish.

Put your agreement in writing, and make two copies, one for each of you to sign. This can prevent a lot of arguments later on.

If there are things you must do yourself – such as supply the washing machine or the special hand-basin from Italy – make sure you're on time. Otherwise he can claim, and fairly, that you held him up. His time is worth money, and you'll eventually have to pay for it.

Remember that plumbers have 'cash flow' problems
Be prepared to pay about one-third of the total cost of the

job before your plumber begins work. This will be needed for materials, petrol, sandwiches, etc. If the job is to take longer than a week, he will need another one-third about mid-way to completion, in order to pay wages and running costs. When the work is finished and tested to your satisfaction and his, settle the remaining one-third and any extras. Make sure he's done *all* the work. If you're having central heating installed, you should normally retain 10% of the final payment until after the end of the next heating season.

Treat your plumber with respect Keep the relationship on a firm but friendly footing. If you're at home when the work is in progress, offer your plumber an occasional cup of tea, or lunch.

Be prepared to clean up afterwards It's all very well to say that a good workman cleans up after himself. He will, if you insist, but you will pay for it.

3 Clothing, materials and tools

Clothing

There are two main types of work you will be doing; each will be somewhat easier if you are appropriately dressed for it.

For connecting copper or P.V.C. tube, brass fittings and taps, wear your second-best pair of jeans and an old shirt. Roll up your sleeves; you'll be washing your hands a lot (use Swarfega or any other proprietary hand cleanser).

For builders' work, such as making holes through walls, digging up plaster, and any sort of carpentry, try to find baggy, comfortable trousers. Or best of all, boiler suits or dungarees.

Don't wear skirts or shorts (the skin on your knees gets torn when you kneel in the rubble) or tight jeans (they'll fold in creases at the back of your knees and cause severe discomfort after a few hours' work). If you monkey about in attics, or if you open up old chimneys, there'll be dust and/or soot. So, generally speaking, the older the clothes you choose the better off you'll be. You can always throw them away when the job is done.

Buying materials and tools

One of the reasons for doing it yourself is to save money. Therefore, if at all possible, arrange to buy your materials and tools at a discount. It is very common to be able to buy at a discount of 10% to 15%; on occasion, up to 33% discount is available to tradesmen.

Try buying your building supplies at lunchtime or mid-morning during the week; slip a boiler suit over your office clothes, and you may be taken for a tradesman. To make your 'image' even more believable, take a standard, printed Supply Book (from a stationer's) with you. Write your purchases in the book as you go. When everything's down, hand over the book and ask the shop assistant to fill in the prices, discount and tax, and to total up. Pay in cash, and off you go.

Where to shop

Do-it-yourself shops These are useful because they sell small quantities, they're open on Saturday afternoons or even Sundays, and because the shop owner will often be able to give the amateur the sort of advice he needs most. He's used to sorting out the problems of people with two left hands. But these shops rarely give discounts.

Multiple chainstore, or modern builders' merchant Here, a large range of articles will be laid out as in a supermarket, so that you can see what's available without having to ask for it (useful, especially, if you don't know what it's called – and terminology can vary!) Usually as expensive as speciality do-it-yourself shops, though they often give good discounts to professionals. So wear your boiler suit, take your order book, and be prepared to argue. Some will give discounts only after you've opened an account, though, and this usually means bank and trade references and a two months' wait while the paper mountain moves. If they're that exclusive, go elsewhere. But do watch for bulk purchases, or clearance sales, etc.

Ye Olde Builders' Merchante He will have everything . . . somewhere. But you will need to know the name of what you need, and how to use it. There will often be other people in the shop who will give you good advice as they wait their turn, though the chap behind the counter only sells things. Not always the cheapest, but usually willing to give a builder's discount to almost anyone, without references. Another plus: these shops often have old stock lying around for two or three years, which they're still selling at the old price.

In general, shop around, look for sales (especially useful when buying big items such as basins, sinks, baths and taps) and auctions. Don't be afraid to open your mouth and *ask* for a discount before you buy. Try to buy everything you need in one go – it looks more impressive, and gets you that discount. If you need additional extras later on, the same shop won't hesitate to give you a discount the second time you go there.

4 Getting to know your plumbing system

You'll need pencil and paper, Sellotape, a torch, and perhaps your old clothes to protect you as you explore your house from cellar to attic.

The cold water system

Running along the road outside the house will be the water mains, which supply water to all the houses in the street. As the mains pass each house, a side pipe will lead from it, through which water is delivered to each one. Where the side pipe crosses the boundary between house and street, there should be an external stop-cock. (See Section 7.)

Normally all you can see of the stop-cock is a small, metal

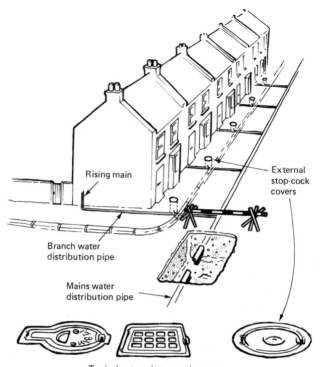

Rising main

External stop-cock covers

Branch water distribution pipe

Mains water distribution pipe

Typical external stop-cock covers

Fig. 1

cover in the pavement. It may be round, square, or shaped a bit like a giant keyhole. This is the Water Board's stop-cock and only the Water Board can turn the water off here.

From the external stop-cock, the water-supply pipe runs on until it enters the house. Immediately inside, it should be fitted with a second stop-cock, which is called the main stop-cock. This is what you must turn on and off when you have to do any work to the plumbing inside the house. It may be in your cellar, or under a wooden hatch below the front door mat, or under the steps to the front door.

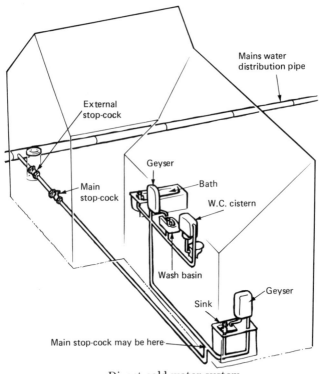

Direct cold water system

Fig. 2

Sometimes there is no main stop-cock. In this case, only the Water Board can turn off your water – which can be most inconvenient if you wish to renew a mere tap-washer. If your house has no main stop-cock, it may be advisable to have one fitted (Sections 2 and 6).

If your house has neither basement, cellar or main stop-cock, look for the water-supply pipe in or near the kitchen. The pipe will rise straight up from the floor and run directly to the tap over the kitchen sink. Follow the pipe from the cold-water tap over the sink until it disappears, either into an outside wall, or into the ground floor.

From the main stop-cock onwards, the water-supply pipe is called the rising main, because it rises from the lowest level of the house up to the attic. It first supplies fresh mains water to your kitchen tap. From there, it does one of two things: either it supplies each tap in the house with fresh mains water (the direct system), or it goes to the attic and empties into a large storage tank (the indirect system), which in turn supplies taps over basin and bath with lower-pressure water.

If there is no main stop-cock, and/or the house is a small cottage in a back street of a large city, chances are good that it uses the direct system – especially if it was built before 1900 or so, and hasn't been modified since. To check this, switch off the main stop-cock (clockwise) if there is one; check to see if any water comes out of the tap over bath or washbasin. If not, then your system is direct. If there isn't a main stop-cock, you will have to arrange to check this by having the external stop-cock turned off in the street.

If you have a direct system, you will have mains water coming out of every tap; the rising main will wind its way up to the highest water appliance, and then it will stop.

An indirect system is more complicated. To follow it, begin by looking at the cistern in the attic. When you wriggle up there (probably through a hatch in the ceiling over the stairwell, or perhaps in one of the bedrooms) take care to step only on the joists – the large pieces of wood spanning from one wall to the other. Joists stick up a little from what appears to be a very rough and dusty attic floor, but in fact that floor is the ceiling of the room below it. If you step between the joists, you go through the ceiling.

Somewhere in the attic, you will see a large storage tank. Usually you will hear the hiss of water entering it. There will be several pipes going into the tank, one of which will be the end of the rising main, which finishes up near the top of the tank. Another pipe, also near the tank's top – but slightly thicker – will make its way to an outside wall, or else it will disappear into the angle between the roof and the 'floor' of the attic; this is the overflow pipe. The pipes

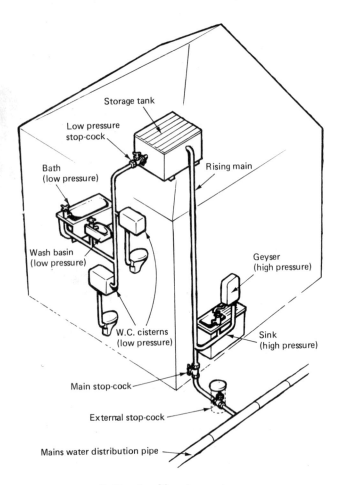

Indirect cold water system

Fig. 3

near the bottom of the cistern will be the distribution pipes to the hot and cold taps in the house; these should be fitted with stop-cocks, although this has not always been done. (See Section 7.)

The storage tank holds any amount up to 100 gallons (450 litres) of water. It is this water which comes out of the cold taps over basin or bath, and which runs into the w.c. cistern. As the water runs out of the tap over the basin, it is replaced by new water coming into the cistern. This system ensures that there is always enough water stored inside your house to flush the w.c. and to wash your hands afterwards, even if the Water Board is digging up your road and turns off *all* the water for half a day or so. If your house doesn't have a storage tank, consider having one fitted – though it's not essential, nor is it a legal requirement, if your house was built without one. (See Section 15.)

From the attic storage tank, the water runs down the distribution pipes until it comes to all the taps and w.c. cisterns in the house. Since these are not supplied by water under mains pressure – but only by water under the pressure of gravity – pipes need to be larger, to allow enough water to flow so as not to end up as a trickle at the taps. It is the vertical distance between the water level of the tank in the attic and the tap being used (called the water head) which determines how much pressure 'squirts' the water out of the tap.

The hot water system

Your house may also have a hot water system. All the pipes may look the same, and the only way to tell them apart is to follow them through rooms, walls and floors until you come to a cold tap or a hot tap. You may have to follow a pipe in both directions before you find the tap. When you have determined what any particular pipe does in any room, label it with a piece of paper stuck on with Sellotape to show whether it is cold or hot, and in which direction the tap is placed.

Trace the pipes back from all hot taps to the point where they enter a large copper cylinder, often found in the airing cupboard. This cylinder should have an insulating jacket.

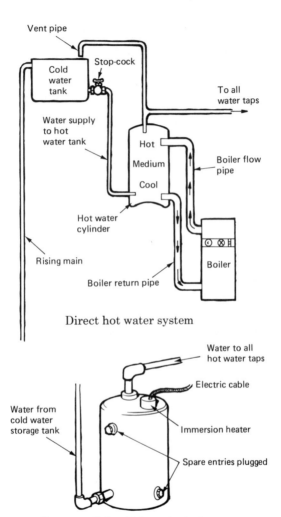

Direct hot water system

Hot water cylinder with immersion heater

Fig. 4

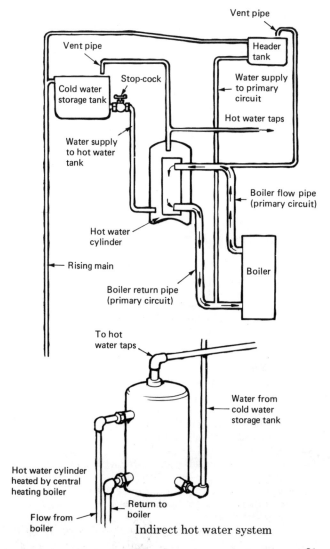

Vent pipe

Vent pipe

Header tank

Cold water storage tank

Stop-cock

Water supply to primary circuit

Hot water taps

Water supply to hot water tank

Boiler flow pipe (primary circuit)

Hot water cylinder

Rising main

Boiler

Boiler return pipe (primary circuit)

To hot water taps

Water from cold water storage tank

Hot water cylinder heated by central heating boiler

Flow from boiler

Return to boiler

Indirect hot water system

From this cylinder, there will usually be two pipes which make their way to the storage tank in the attic. One will be a distribution pipe *from* the cold water storage tank *to* the hot water storage tank (as the copper cylinder is called). Each time you run a hot tap, cold water from the attic tank will come into the copper cylinder to be heated. The second pipe, running from the *top* of the copper cylinder, goes up to the attic and finishes about 6 in. (15 cm) above the cold water storage tank, but not actually touching it. When water in the copper cylinder gets too hot, it expands. It runs up this second pipe to drop into the attic tank. Without such a pipe, the copper cylinder might explode if overheated. A branch from this second pipe will also be the hot water distribution pipe to all your hot taps.

If the copper cylinder is heated only by an immersion heater, there will be no further pipes coming into it. You'll simply see a thick electric cable which disappears into a large plastic box near the top of it, and there will be a large switch nearby.

If the cylinder is heated by a gas- or oil-fired boiler (or by coal) then there will be two further pipes coming into the cylinder, making four in all. The two additional pipes are the flow and return, to and from the central-heating boiler. They'll be somewhat larger than the other two, which is how you can recognise them without following them all the way to the boiler.

Drainage

Now that hot or cold water is supplied to each appliance, all of them must be drained. If you look under a sink, you'll see a complex twist of tubes and round shapes immediately below the bowl. This is called the trap. Running away from it is a straight length of pipe (usually horizontal) called the waste-pipe.

The water seal formed by the water settling in the U-trap is fitted to all sinks, basins, baths and w.c.s. The water in the w.c. pan is only the top of this water seal; what T.V. commercials refer to as 'the bend' is in fact this same U-trap, only much larger.

From the U-trap, waste water runs down to the drains. In

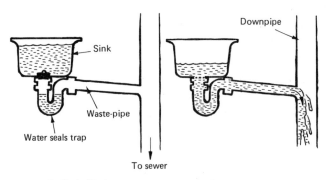

1. Basin filled

2. Pull the plug out, water rushes down trap and into downpipe

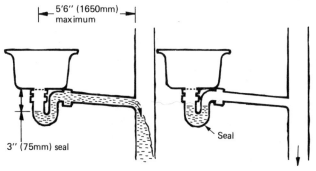

3. Last of water rushing down waste-pipe tries to suck all water out of trap but can't.

4. Permanent seal stops smells coming up waste-pipe

How the seal in the trap works

Fig. 5

order to reach them, especially if the waste-pipe is on the first floor, it must first come to a downpipe or soil-stack. Downpipes or soil-stacks are vertical pipes, from 2 in. (5 cm) to 6 in. (15 cm) in diameter which run from the top of the house to the bottom, and then into the main drain.

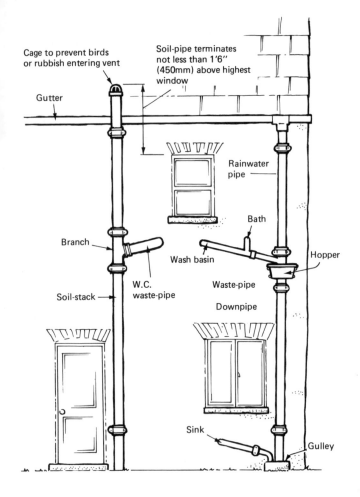

Two-pipe system

Fig. 6

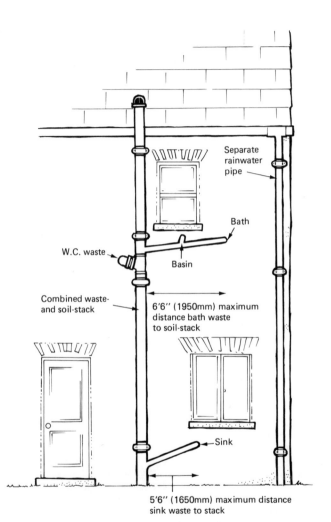

Separate rainwater pipe

Bath

W.C. waste

Basin

Combined waste- and soil-stack

6'6'' (1950mm) maximum distance bath waste to soil-stack

Sink

5'6'' (1650mm) maximum distance sink waste to stack

One-pipe system

Depending upon the age of the house, it will be served either by a soil-stack *and* a downpipe (two-pipe system) or by one single, large pipe which takes waste-pipes both from basins and from w.c.s (combined, or one-pipe system). The one-pipe system may be installed either inside or outside the house, but the two-pipe system is *always* outside.

Identify (from outside the house) your bathroom and kitchen windows. If the plumbing is external, you will notice a small pipe, about $1\frac{1}{2}$ in. (4 cm) in diameter, which comes out of the wall and slopes down to a larger pipe which runs vertically. If there is no other pipe anywhere on the face of the building, the house is equipped with a one-pipe system. If there is a second, somewhat larger pipe taking side pipes from the wall near your w.c., then you have a two-pipe system. If there are no pipes at all on any outside wall, there is a one-pipe system inside the building.

In a two-pipe system, waste-pipes will run from the sink and wash basin into the smaller downpipe or into a special Y-branch. On the ground floor, the waste-pipe may run into a gulley. The waste-pipe from the w.c. will be much larger, usually 4 in. (10 cm) in diameter, and will run into the soil-stack, which will also be 4 in. (10 cm) in diameter. In a one-pipe system, wastes from wash basins and sinks as well as from w.c.s will enter the same pipe directly, with no hoppers or gullies.

At the foot of the building, below the ground, the soil-stack will change to a horizontal direction, and make its way to the main sewer in the street. Before it gets there, it must join with any other downpipes or soil-stacks (if it is a large house, or a block of flats). There will be a manhole at the junction.

A manhole is covered by cast-iron, the cover usually about 18 in. × 27 in. (45 × 68 cm) with a ribbed surface and two half-round indentations at each end, so that you can pick up the cover. If you were to lift it, you'd see two or more tunnels of about 4 or 5 in. diameter (10 to 12.5 cm) converging, and one tunnel going down to the main sewer. The manhole is there to allow someone (but not you, not in this book anyway) to get inside to clean out the drains, if they get blocked.

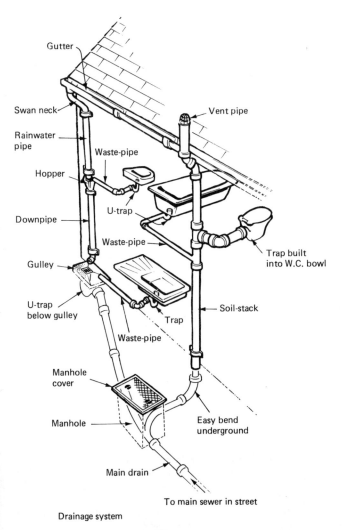

Gutter

Swan neck

Rainwater pipe

Waste-pipe

Vent pipe

Hopper

U-trap

Downpipe

Waste-pipe

Trap built into W.C. bowl

Gulley

U-trap below gulley

Waste-pipe

Trap

Soil-stack

Manhole cover

Manhole

Easy bend underground

Main drain

To main sewer in street

Drainage system

Fig. 7

The rain which falls on your house runs down the roof to collect into gutters, half-round (or sometimes square) lengths of plastic or metal which are fitted to the edge of the roof, all around the house. Water collects in these, and the gutters are arranged to slope slightly in one direction along their length. The water flows down the length, and at the gutter's lowest point there is a downpipe which takes the water to the main sewer. In a two-pipe system, the rainwater comes down in the same pipe as the bath water; in a one-pipe system, rainwater requires its own downpipes.

5 Tools

The tools you will need to tackle the jobs described in this book come in three categories: emergency; basic; useful but not essential, useful for one job only, or useful if already owned.

Emergency

Electrical insulation tape of the modern thick P.V.C. type, is useful for bandaging leaking pipes and joints.

Spiral curtain wire (as used for holding up lace curtains) is good for wriggling down waste-pipes to get them clear. A wire coat-hanger, cut with a pair of pliers and straightened out, can also be used to get to awkward places.

The adjustable wrench is used for loosening and tightening pipes and nuts and is described in the next section.

A plumber's mate is made up of a wooden handle with a rubber cup at the end; sizes vary from 10–18 in. long (250–450 mm) with a cup varying between 4 and 6 in. (100–150 mm) in diameter. Vigorous pumping action with the handle while the cup is held over the blocked sink waste can clear the blockage. It is also known as a plunger or a suction cap.

PFTE tape is a thin tape that helps to seal leaking screwed joints.

The use of these tools and materials in the emergency section is set out in greater detail in the sections on blocked pipes and leaks (Sections 9 and 10).

Basic tools

An adjustable wrench can grip and turn any pipe or nut; it can exert enough force to snap off a bolt or twist a thin-walled pipe. It also leaves teeth marks if a lot of force has to be used – wrap the nut or pipe in cloth or Sellotape if you don't want scratches.

To use it, open the jaws as with an ordinary set of pliers; slip the jaws over the nut or pipe and squeeze the handles together with one hand while pushing the outer handle in the direction you want the assembly to move. Always arrange the wrench so that you push in the direction you want the nut or pipe to move – never pull. For very stiff nuts or pipes, use a 'cheater', which is a long length of pipe slipped over the outer handle, as in *d* (Fig. 8).

To adjust the jaws, undo the pivot bolt (anti-clockwise) and pull it right out. You find that the inner handle slides freely up or down inside the outer handle. Push back the pivot bolt, and slide the inner handle till the bolt engages one of the four or five holes – pick a hole which gives you an opening between the jaws that will allow you to grip the nut or pipe firmly with one hand. Screw back the pivot bolt.

Adjustable pliers These won't grip a pipe as tightly as the adjustable wrench, but have a slimmer body and much bigger jaws. They can therefore take much larger nuts, and slip into more constricted spaces. (Fig. 9).

To adjust the jaws, the handles are moved as far apart as possible, after which you'll find that the arm with the serrated slot can be moved with regard to the other arm in such a way that the jaws are further or nearer.

To use the pliers, adjust the opening of the jaws till you can grip the nut firmly using one hand only. Grip the handles as tight as possible, and you can turn the pliers either way, i.e. either pushing or pulling.

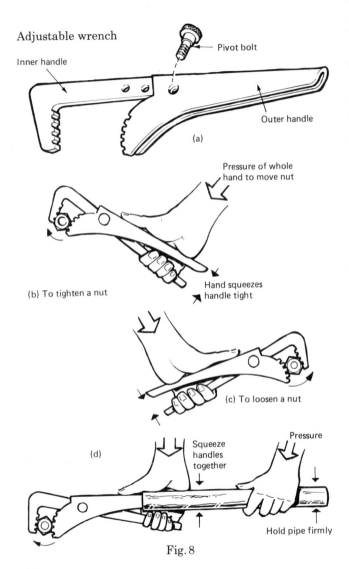

Adjustable wrench

Inner handle

Pivot bolt

Outer handle

(a)

Pressure of whole hand to move nut

(b) To tighten a nut

Hand squeezes handle tight

(c) To loosen a nut

(d)

Squeeze handles together

Pressure

Hold pipe firmly

Fig. 8

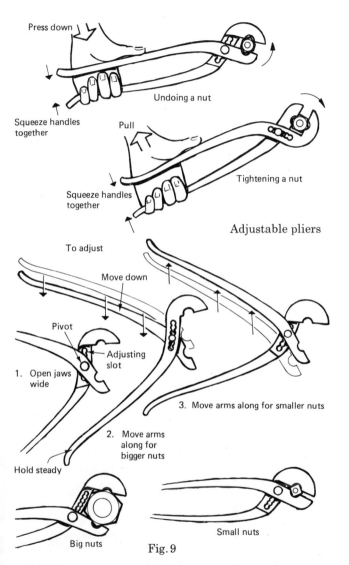

Press down

Undoing a nut

Squeeze handles together

Pull

Tightening a nut

Squeeze handles together

Adjustable pliers

To adjust

Move down

Pivot

Adjusting slot

1. Open jaws wide

Hold steady

2. Move arms along for bigger nuts

3. Move arms along for smaller nuts

Big nuts

Small nuts

Fig. 9

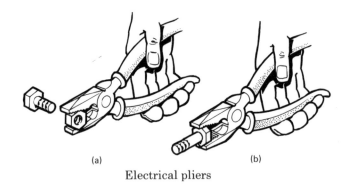

(a)   (b)

Electrical pliers

Fig. 10

Electrical pliers These are useful for gripping small nuts or bolts; they can also be used for cutting wire. They will grip a nut more firmly if held as in *a*, but can also be used as in *b* – sometimes useful in confined spaces, although it doesn't grip the nut as tightly.

Blowlamp Nowadays the only blowlamp used in plumbing is the type fitted with a gas cartridge; to use it simply switch the gas on and light the flame. Remember to keep a spare cartridge for a week-end's work. Remember, too, that the hottest part of the flame is at the tip of the inner blue cone of flame, and it is this part of the flame which should be played over a welded joint.

Large and small screwdrivers All screwdrivers of whatever size are useful at some time. If you're buying new, choose one about 10 in. (250 mm) long, and one small electrical screwdriver. See Section 8 on disassembly for how best to use a screwdriver.

Junior hacksaw This is a very simple type with a very small and thin blade which doesn't last long, so make sure you have spares for the week-end. To use it, put the blade on the place where you want to cut the pipe, and pull it back

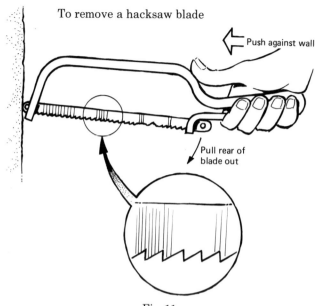

To remove a hacksaw blade

Push against wall

Pull rear of
blade out

Fig. 11

three or four times. You'll have started a small groove, and
now you can push the blade as well as pull, thus making a
simple to-and-fro movement. Don't jerk the blade, or put
too much pressure on it, since it is really not very strong.
Make sure that you don't finish each stroke in the same
position, but vary the length of your stroke; if you don't,
the blade will twist at the point where you stop all the
strokes, and then simply snap.

To replace a blade, hold the saw by the handle, and push
the nose firmly against a wall. You can now pull the rear
end of the blade down; relax the pressure on the frame, and
pull the front end of the blade out of its socket.

Hold the new blade against the light, to see which way
the teeth point. Put the blade into the frame in such a way
that the teeth point forward. Start by slotting the front of
the blade into the nose of the frame; then hold the nose

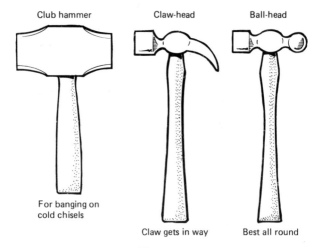

Club hammer

Claw-head

Ball-head

For banging on
cold chisels

Claw gets in way

Best all round

Fig. 12

against the wall, push hard, and with your other hand slip the rear of the blade into the slot at the rear of the frame.

Hammer Any hammer will do, but for light work the ball-headed engineer's hammer is best, and for heavy work, such as making holes in walls, a club hammer is best. A claw hammer takes up more room, which is a disadvantage when working in confined spaces. Use the side of the hammer if there's no room to swing it in the normal way.

Chisel In plumbing cold (or steel) chisels are used to make holes in or through walls. Choose a short one with a broadish blade for cutting grooves along the wall if you want to bury pipes; a long one, at least 11 in. (300 mm) long for going through brick walls, and a medium one for starting holes to go through walls. Keep the points sharp – you can buy a small arbor that fits on to the end of an electric drill to do this.

Use a cold chisel in short spells of not more than a minute at a time – your arms will get tired if you use it longer.

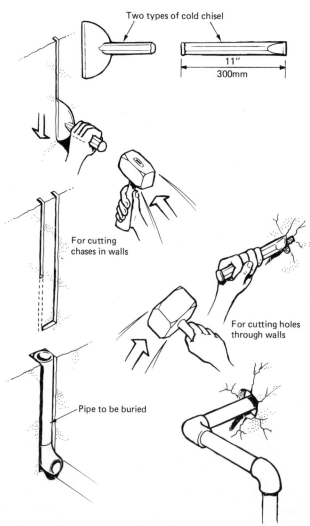

Two types of cold chisel

11"
300mm

For cutting
chases in walls

For cutting holes
through walls

Pipe to be buried

Fig. 13

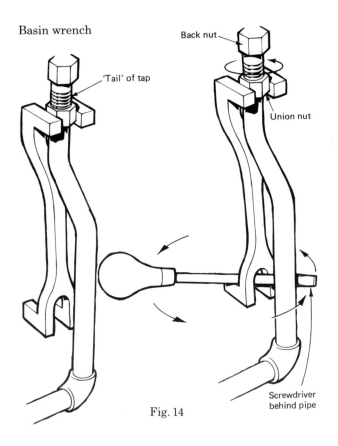

Basin wrench

'Tail' of tap

Back nut

Union nut

Screwdriver
behind pipe

Fig. 14

Tools that are inessential, used for one job only, or nice if you've got them already

Basin Wrench This is a special type of spanner where the jaws of the spanner are at right-angles to the shaft. It is used to tighten the back nut round the tail of a tap if the bath holding the tap cannot be turned upside down or moved in any other way.

Pipe cutter

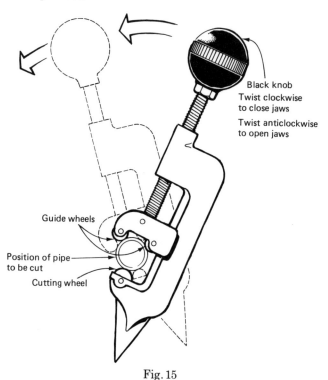

Black knob
Twist clockwise
to close jaws

Twist anticlockwise
to open jaws

Guide wheels

Position of pipe
to be cut

Cutting wheel

Fig. 15

To use it, slip the jaws round the *back* of the pipe leading up to the underside of the tap, and push the wrench up, wriggling a little bit all the while, till the jaws engage the nut. Slip a screwdriver or short metal bar in the lower jaw to act as a lever to turn the wrench round. Slowly and firmly revolve the wrench till the back nut is loose.

Pipe cutter This is a device for making perfect and smooth cuts in copper tube. To use it, twist the black knob anti-

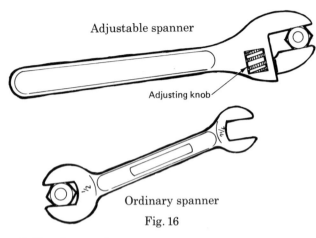

Adjustable spanner

Adjusting knob

Ordinary spanner

Fig. 16

Stillson's wrench

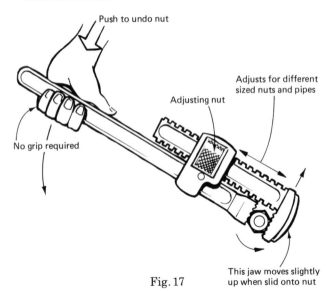

Push to undo nut

Adjusts for different sized nuts and pipes

Adjusting nut

No grip required

This jaw moves slightly up when slid onto nut

Fig. 17

clockwise till the jaws are far enough apart to slip over the tube. Now turn the black knob clockwise till the jaws grip the tube. Revolve the whole device once round the tube, turn the black knob once round clockwise, and revolve again. Keep repeating till the tube is severed.

Stillson's wrench This is an adjusting wrench for gripping nuts and pipes – it does the same work as the one described earlier on, but is less easy to use, and costs more. If you have one already, you'll know how to use it, and won't need to buy a second adjustable wrench.

Adjustable spanner This will take any size nut (within the limit of its adjustment) and turn it round without leaving teeth marks. The adjusting mechanism tends to wear out with time, so that the adjustable spanner doesn't grip the nut very firmly any more.

Ordinary spanners Each spanner will only fit one, or at the most two, sizes of nut; also the big spanners needed for most plumbing work cost a great deal. If you have a set already, always check to see if any fit the plumbing fittings.

6 Materials

In old-fashioned houses (pre-1920), lead water pipes are very common. Less common is gas barrel pipe.

Lead pipe used in existing plumbing systems will almost certainly be badly worn by now, and may leak; in that case it will *have* to be renewed. In any case, new fittings may be desired for additional water appliances in a house served by a lead pipe plumbing system. Where gas barrel pipe has been used for water, it will almost certainly rust eventually and require replacement.

Any modern extension or change to existing plumbing systems, whether they are lead pipe or gas barrel, will be made in copper tube. Whether your system is lead pipe or gas barrel, you will need to know how to connect it to copper.

1. Plumber puts in stop-cock

2. 1st week-end:
   renew downstairs cold water

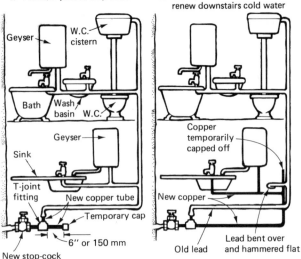

Geyser

W.C. cistern

Bath Wash basin W.C.

Geyser

Sink

T-joint fitting

New copper tube

Temporary cap

6" or 150 mm

New stop-cock

Copper temporarily capped off

New copper

Old lead

Lead bent over and hammered flat

3. 2nd/3rd week-end: progressive stages

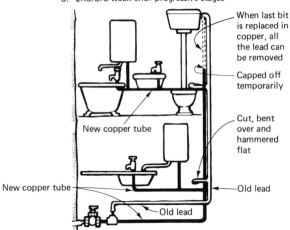

When last bit is replaced in copper, all the lead can be removed

Capped off temporarily

Cut, bent over and hammered flat

New copper tube

Old lead

New copper tube

Old lead

Stages in converting an old lead to new copper system

Fig. 18

Lead pipe to copper

There are two possibilities to consider when joining existing lead pipe systems to copper.

New copper tube and/or fittings can be added as you go along, piecemeal, as leaks develop or as new water appliances are required in the house. This is not a job for absolute beginners. Joining copper pipe to lead pipe requires the skill of a professional plumber with a lot of practice in the techniques involved.

The simpler (and to my way of thinking, better) plan is to renew the entire existing lead pipe plumbing system with copper tube. This will require the services of a professional plumber once only. It is done as follows:

Begin by checking your existing plumbing installation to determine whether or not you have a main stop-cock. (See Sections 4 and 7.)

Where a main stop cock does exist, hire a plumber to renew it completely, and to replace part of the tail in copper. This procedure will always involve asking the Water Board to turn off your water supply at the external stop-cock.

If your house does not have a main stop-cock, ask the plumber to install one, complete with a copper tail, so that you end up with the arrangement shown in the diagram.

You can then proceed to install your own T-junctions in copper (see Section 6) as and when it is necessary or convenient to do so. Thus you can renew the lead pipe system gradually, in easy stages (rather than having the entire house upside-down for months). Furthermore, you will never have to join lead pipe to copper. Fig. 18 illustrates the further stages of the replacement of lead pipe by copper tube.

Gas barrel to copper

Gas barrel is a thick-walled mild steel tube, threaded at both ends to take further tubes, T-joints, elbows and bends, straight joints or whatever else might be required. It was normally bought in various standard lengths, already fitted out with screw threads at each end.

Gas barrel to copper

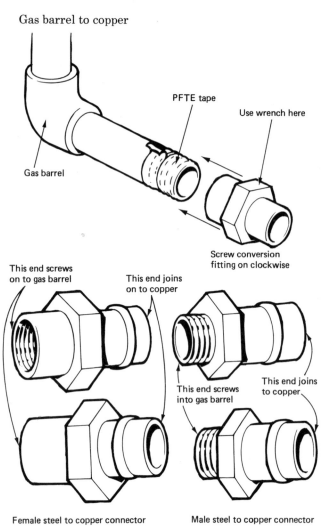

PFTE tape

Use wrench here

Gas barrel

Screw conversion
fitting on clockwise

This end screws
on to gas barrel

This end joins
on to copper

This end screws
into gas barrel

This end joins
to copper

Female steel to copper connector

Male steel to copper connector

Fig. 19

To convert gas barrel to copper, you must first take all the gas barrel pipework away, section by section, until you have reached the point at which you wish to connect copper tube. This requires considerable physical strength, and you may find you need the help of someone strong(er) when you attempt to unscrew gas barrel sections.

You will then have to purchase the appropriate brass-barrel-to-copper fitting. This will screw on to gas barrel on one side, and connect to copper on the other, as shown in Fig. 19.

To choose the appropriate conversion fitting, begin by measuring the inside, as well as the outside, of the gas barrel. Better still, take the last of the gas barrel you just unscrewed (if it isn't too long) to the ironmonger. Buy the fitting, and also a reel of PFTE tape.

Wind the PFTE tape once around any exposed thread of either the gas barrel or the brass fitting, depending on whether the gas barrel is a male spigot or a female socket. Screw the brass fitting to finger tightness, and then use a single wrench to turn it as tight as it will go. When it is tight, put *all* your strength on the wrench by putting the full weight of your body on the handle, and tighten the fitting a bit more. It is now ready for connecting to copper tube, as described below.

Copper tube

Copper tube comes in 12 mm ($\frac{3}{8}$ in.), 15 mm ($\frac{1}{2}$ in.), and 22 mm ($\frac{3}{4}$ in.) diameters. It is strong, and cannot be bent to go around corners unless you have special equipment. But if a very long section is accidentally jammed in a doorway or between floor and wall, it will bend and crease. The crease will not come out, and the area around the fold will be useless, so handle long sections carefully.

15 mm ($\frac{1}{2}$ in.) copper tube is the standard size used currently in the U.K.; 22 mm ($\frac{3}{4}$ in.) is used to supply hot and cold water to the bath, and for long pipe-runs, especially if more than one tap is supplied from the run.

You will rarely need to bend copper tube, since you can buy elbows. But bending the tube is sometimes necessary for the last fiddly bit between tap and water carcase.

Fittings

Fittings are the connectors between pipes or tubes which allow for extending the network, or for changing direction. Fittings are usually of brass, whether they are used for copper, steel or lead pipework, or any metal combination. There are three main types of joint you will probably see at one time or another, as shown in Fig. 20.

Straight connector joint fittings are used to join two pieces of the same material in order to form a very long, straight run, when several short sections of the same material are joined together, to join two different types of material, and when a special type of straight connector called a union joint is used to join the last of the water-pipe to an appliance.

A variation of the union joint is the swivelling joint, used to connect the tail of a tap to the last of the water pipe.

Elbow, or bend, fittings are used to go around corners. The elbow has a small radius, the bend a larger one. You can use the elbow or bend either to join two pieces of the same material, or to join two different materials.

T-fittings take a branch from main to side pipes. These normally feature the same material on all three openings, and take the same pipe-size throughout.

Reducing fittings will enable you to connect two different-sized pipes or tubes, and are sometimes incorporated in gas-barrel-to-copper connectors. When buying reducing fittings, or copper-to-steel connectors, you must measure the inside and the outside of the pipe or tube which is to be connected. You must also check to see if it is a male or a female connection. It is best to make a small sketch of the pipe-end before going to the ironmonger.

When you have a run of pipe at the end of which is (or used to be, or will be) a tap or a fitting – but which at present needs to be stopped off – use a cap or a plug, depending on whether you have a male or female terminal.

Jointing work

There are two methods of joining copper to copper, using connector, straight, or elbow fittings:

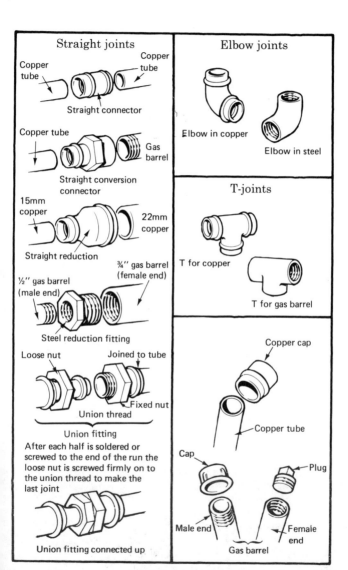

Straight joints

Copper tube — Copper tube

Straight connector

Copper tube — Gas barrel

Straight conversion connector

15mm copper — 22mm copper

Straight reduction

½" gas barrel (male end) — ¾" gas barrel (female end)

Steel reduction fitting

Loose nut — Joined to tube

Fixed nut

Union thread

Union fitting

After each half is soldered or screwed to the end of the run the loose nut is screwed firmly on to the union thread to make the last joint

Union fitting connected up

Elbow joints

Elbow in copper

Elbow in steel

T-joints

T for copper

T for gas barrel

Copper cap

Copper tube

Cap — Plug

Male end — Female end

Gas barrel

Fig. 20

45

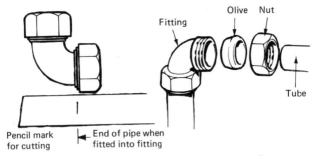

Pencil mark for cutting | ◄── End of pipe when fitted into fitting

1. Cut pipe to size

2. Undo the nut, slip nut then olive on to pipe fitting

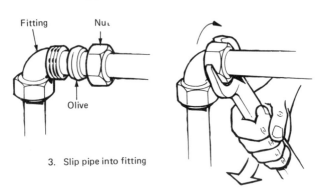

3. Slip pipe into fitting

4. Screw on nut and tighten

Installing compression fittings

Fig. 21

Compression fittings which require the use of two wrenches, or adjustable spanners.

Cut the tube to the required length, leaving an extra piece to go into the fitting. Use a pipe-cutter if you can; otherwise, use a hacksaw. Clean off the end of the tube to remove burrs, using a small file.

Pick up the compression fitting and unscrew the nut.

Look for the small brass ring which sits inside. This is called an olive. It has a sharper edge, and a blunter edge. Slip the nut over the tube, then slip the olive over the tube with the blunter edge towards the nut. Now slip the tube into the compression fitting, and push the olive towards the fitting and into it. Finally, push the nut toward the screw-thread portion of the fitting and turn the nut until it becomes difficult to tighten any further. You'll hear creaks and groans from the olive when she's had enough.

There is no way to tell if the joint leaks until you turn on the water. If it does, try turning the nut a bit tighter – until the leaking stops. If it doesn't, however hard you turn the nut, switch the water off, undo the whole fitting, and wrap PFTE tape around the pipe and olive before doing up the fitting really tightly. That should stop the leak.

The main advantage of compression fittings is that you can undo them easily, which is useful if you have to rearrange the system at a later date. The olive won't come off, and is usually crimped into the tube. You will have to shorten the tube by cutting off the part which holds the olive. You will also have to buy a new olive.

Capillary fittings which require the use of a blowlamp.

Cut the tube to the required length, leaving an extra bit to enter the fitting. Take care to leave the tube feeling smooth by cleaning off all rough edges and burrs.

Wrap steel wool around the end of the copper tube, twist it round and round to clean the copper until it shines. Clean the inside of the copper fitting which will take the tube. Make a pencil mark about $\frac{1}{2}$ in. (1 cm) from the end of the tube, then spread flux paste all over the end of the tube, and inside the fitting.

Slip the tube into the fitting, check to be sure it enters smoothly and all the way to the shoulder inside the fitting. Check this against the pencil mark you made earlier.

Now prepare, in a similar fashion, all the tubes which are to enter the fitting. When you heat the junction between tube and fitting, it will be nearly impossible to avoid heating all other parts of the fitting, since it is such a small piece of metal. If you wish to leave one opening free, then cut 6 in. (15 cm) or so of copper tube, leave it uncleaned and unfluxed, and push it into the opening to be left clear; wrap a piece of wet cloth around it before you use the blowlamp.

Light the blowlamp and play the flame over the fitting, where the tube enters it. After about 30 to 45 seconds, the solder will start oozing out of the join. Hold the flame for another 2 seconds, and then remove it completely. If you have cleaned tube and fitting properly, and applied flux, then the solder will appear as a silver ring all around the copper tube where it joins the fitting. If it doesn't go all the way around, heat the fitting a bit more – perhaps about 10 seconds. The trick is not to heat the solder too much, since it will run to the bottom of the tube and leave you with a leaking joint. This doesn't happen easily, so don't worry too much. Just be sure to heat the joint only enough to finish with a nice, complete, silver ring around the tube.

Repeat the process for the other parts of the fitting, bearing in mind that these will already be hot from the earlier heating. You will need to heat each of the others for a few seconds only.

Capillary fittings are smaller than compression fittings, and require less room between wall and copper tube (and thus are used for gas, as well as water). Their main disadvantage is that if they are not fixed properly the first time, they are never fixed properly. If they leak, you cannot fix them. You have to remove the whole fitting, clean the copper tube with steel wool until it shines, apply flux paste, and use a new fitting. Make sure that all pipes are dry – really dry, inside and out – before trying again.

To remove old fittings, heat the joint with a blowlamp while gripping the fitting with an adjustable grip; as the solder is heated to flowing point, the whole fitting can be twisted loose and off the copper tube. This is useful to know if you want to rearrange the plumbing system. Using this method, you can swivel the joint into another direction. But you cannot then take it off and put it back on again.

Plastic fittings

Plastic is used mainly for waste-pipes, and all very modern systems employ plastic waste-pipes and traps. If you have

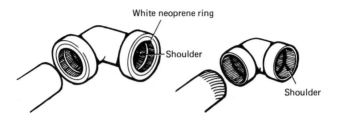

White neoprene ring

Shoulder

Shoulder

Push hard, twist and wriggle till
tube squeezes past neoprene ring
to sit firmly against shoulder

Coat end of tube with special
solvent glue and push into
fitting, twisting the whole to
ensure even distribution

Push-fitting

Solvent-weld fitting

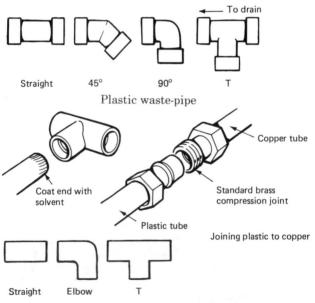

To drain →

Straight   45°   90°   T

Plastic waste-pipe

Coat end with
solvent

Copper tube

Standard brass
compression joint

Plastic tube

Joining plastic to copper

Straight   Elbow   T

Plastic water pipe (cold)

Fig. 22

existing lead, copper (or other metal) waste-pipes, it is perfectly possible and very easy to join these to plastic. Your plumber's merchant will sell you the special fittings if you give him the size of the old pipe. Since plastic waste and fittings are (comparatively) very cheap, it is sensible to replace the entire lot in plastic, unless the existing runs are very long.

Plastic pipe is bought in lengths of 3 metres, though you can – for an inflated extra price – buy shorter lengths. The joints are straight, elbow and T, though with subtle variations. The elbow comes in three angles, $22\frac{1}{2}°$, $45°$, and $90°$; the T features a 'swept' bend which must be fitted so that the streamlined junction faces the direction from which the water flows.

Push-on fittings can be disassembled whenever anything gets really stuck, or if you wish to modify the system. This is useful, since you need buy only a short length of new waste-pipe, and existing fittings can be re-used.

Solvent-welded fittings are slightly less bulgy at the joints, and they are slightly cheaper. But they cannot be taken apart, nor can they be re-used. They are the better choice on outside installations, due to the danger of vandalism, etc.

Plastic high-pressure water pipe

Although not in common use (some Water Boards either disapprove, or do not encourage its use) plastic high-pressure water pipe is available. It can only be used for *cold* water pipes.

The pipe comes in long rolls, and can be cut off into required lengths with a tenon, or junior hacksaw. The fittings are the usual T, elbow and straight couplings, and are of the solvent-weld type. It is cheaper than copper, though you will still need copper for hot water systems.

Related Materials

Putty is used to seal the joint between wastes in sinks and basins, and the actual metal or porcelain of the appliance.

Use ordinary glazing putty, which is available in small drums, if you're going to need a lot. If you need it only for plumbing, buy it in small, plastic squeezy pack form. To use, squeeze or dig some out, knead between your hands for a few minutes until it is perfectly smooth and creamy.

Flux is sold in small tins, and is used to stop the copper tube oxidising when you make capillary joints. Should the copper tube oxidise, then the solder won't flow and 'wet' the copper; hence you'll get a leak at that point. Use the flux after you have cleaned the copper tube with wire wool, and before pushing the copper tube into the fitting. Don't bother to try to make a soldered joint without first using flux; it really won't work.

PFTE tape is a very soft, thin, plastic tape, sold in small rolls like Sellotape. It isn't sticky, though it tends to cling to surfaces just enough to stay in place. It is used to seal the gap between the threads when you screw two items together.
    To use PFTE tape, cut it with scissors (it won't tear) into small lengths, just sufficient to go round the screw-threads of the male spigot once before screwing on to the female socket or nut. The tape is so soft and pliable that it won't offer any resistance to the nut screwing on, but it will stop any water from finding its way out between the threads.

A tip for an emergency If you haven't any PFTE tape with which to make a joint extra-watertight, use the old method; take some sisal string (any unravelled string or even wool will do the trick), unravel the strands, and wind the fibres round the threads. Dip a piece of stick into some gloss paint, and spread it over the string immediately before screwing on the other half of the fitting. Messy, but it works. It can be awfully hard to undo, though, especially after the paint has had twenty years or so to harden.

7 Turning things on and off

Water Systems

Water systems are turned on and off at stop-cocks. Stop-cocks are rather like taps, except that instead of water coming out of the tap, it merely enters another part of the system. Think of stop-cocks like traffic lights or railway signals: when they are open, they allow water to pass into the next section of pipe; when they are shut, the next length of pipe cannot admit water.

Stop-cocks are usually (unfortunately, not always) turned *off* by a *clockwise* motion of the handle, and turned *on* by an *anti-clockwise* motion.

Stop-cocks are sometimes very stiff and hard to turn, especially if they haven't been used for a while. You may have to grip the handle in a pair of adjustable pliers to gain extra leverage. Don't use too much force, though, or else the handle might snap off completely. If you grip the handle by the pliers as in Fig. 23 and *gradually* increase the amount of force you apply – avoiding sudden muscular jerks – you will probably succeed. You might also try applying a bit of penetrating oil at the junction of stem and body.

To turn off the cold water system, you must first find the main stop-cock (see Section 4). If that happens to be outside, say, at the bottom of the garden (as is the case with my house) it may be buried 2 feet (60 cm) below the ground – so that the water is kept above freezing point – and that makes turning it on or off very difficult indeed. I have a special key which came with the house, and which looks like the one in Fig. 24.

If you don't have something like that, you can make your own by sawing a V-notch in a long piece of wood and fitting it with a handle, as in the picture. Whichever you use, remember to turn *clockwise* for *off*, *anti-clockwise* for *on*.

In a direct system (see Section 4), the main stop-cock controls all the taps in the house. If, however, you have a

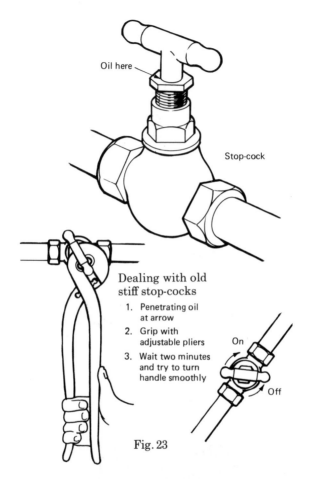

Oil here

Stop-cock

Dealing with old
stiff stop-cocks

1. Penetrating oil
   at arrow

2. Grip with
   adjustable pliers

3. Wait two minutes
   and try to turn
   handle smoothly

On

Off

Fig. 23

storage tank in the attic, then turning off the main stop-cock will only turn off the water supply to the kitchen sink and sometimes to the geyser over the sink or in the bathroom.

Shutting off the main stop-cock will stop the water

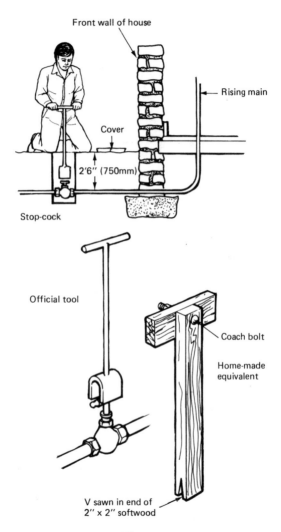

Front wall of house

Rising main

Cover

2'6" (750mm)

Stop-cock

Official tool

Coach bolt

Home-made
equivalent

V sawn in end of
2" x 2" softwood

Fig. 24

coming *into* the storage tank, but it won't stop it coming out. All the cold water taps in the bathroom, as well as the w.c. cistern, will still have 80 or 100 gallons to come. If you haven't got a stop-cock on the distribution pipe immediately next to the main cistern, you will have to open all those taps and allow the entire tank to run dry before doing any work. This takes a very long time, and justifies the installation of a stop-cock on the distribution pipe.

Most systems with a main storage tank do have a stop-cock fitted next to the tank. But if there is more than one distribution pipe, you'll need to trace them in order to find out which is cold-water distribution. Follow it, and you'll find the low-pressure stop-cock. It's not always next to the tank, and it may be near the ceiling of a bedroom (so that you can get at it without climbing into the attic).

If you wish to shut off the hot water system, you must first find the distribution pipe from the cold water tank to the hot water tank. There will be a stop-cock near the foot of the cold water tank, and when this is turned clockwise it will stop more water running into the hot water storage tank. If you now run any hot tap for a few minutes, you'll clear the whole of the hot water system (except for the water still in the hot water storage tank, as shown in Fig. 25).

A word of caution to proud and comfortable owners of central heating systems: the expansion and feed tank on central heating plants using an indirect system will be supplied by water from the rising main. If you intend to shut off the rising main (i.e. to shut off the main stop-cock) for more than a few hours, you should switch off your central heating as well – for reasons of safety.

The same applies to the low-pressure water distribution system if you have a direct central heating system.

If you don't know which system you have, play safe and switch the heating system off whenever you turn off the water. In any case, leave any switches and stop-cocks of your central heating system alone until you understand them. This particular book won't explain them.

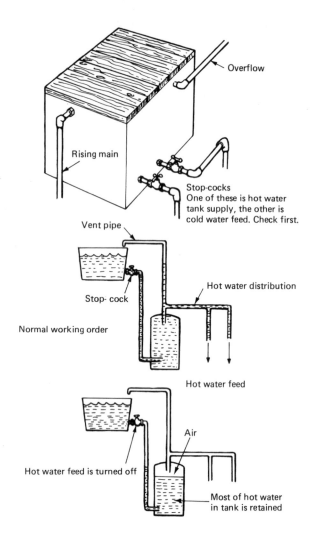

Overflow

Rising main

Stop-cocks
One of these is hot water
tank supply, the other is
cold water feed. Check first.

Vent pipe

Stop-cock

Hot water distribution

Normal working order

Hot water feed

Hot water feed is turned off

Air

Most of hot water
in tank is retained

Fig. 25

8 Assembly, disassembly and re-assembly

Anyone can take a thing to pieces. Putting it together again, though, can be a problem. If you follow my system carefully, with your full attention, you should succeed in re-assembly on nearly every occasion.

First of all, equip yourself with paper, pencil and masking tape (sometimes called draughting tape). Then, as you undo each part of a fitting, make a small sketch of the way it looks *before* you take it apart. Show with arrows where each screw or bolt should go. Use the masking tape to stick a label on each part, so that you can identify exactly where it belongs. It takes time, but it's worth it.

As you unscrew or unbolt each piece, clean it. Use paraffin to remove grease, brushing it on with a stiff brush (which will also remove dirt). This will ensure that the fitting works well when it is put back together again. Furthermore, you'll get your hands dirty only once, and, besides, sticky tape won't stick to dirty surfaces.

Put the pieces into a box so you don't lose them. If you have sufficient space, spread out newspapers on a table and lay out the bolts and pieces in neat rows as you take the fitting apart.

When the fitting is in pieces (or when you have reached the point at which you can repair whatever is wrong) then re-assemble. Now you will use the notes you made when taking the thing apart, as you put the pieces back again in reverse order.

Disassembling fittings will teach you a lot, as will watching other people tackling a job. Make a habit of giving your assemblies a 'dry run'. This means fitting all pipes together, all fittings to pipes or to each other, *without* actually soldering, gluing or tightening any nuts beyond finger tightness. When you're satisfied that the 'dry run' will work, use the glue, the solder, or the spanner.

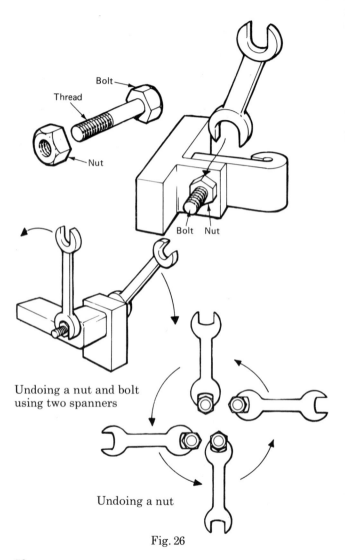

Bolt

Thread

Nut

Bolt Nut

Undoing a nut and bolt
using two spanners

Undoing a nut

Fig. 26

58

General points

All the screw-threads I've ever come across in my plumbing experience have been 'left-hand' screws. That's a technical term which means that screws and nuts must be turned anti-clockwise to undo, and clockwise to tighten.

Fig. 26 illustrates an ordinary nut and bolt which are holding two pieces of metal together.

Assume for a moment that the long, threaded piece of metal (the bolt) remains motionless. If it does, then to loosen the assembly you need to slip a spanner around the nut – with the shaft sticking out to the left – and push down. You could also, if it is easier – and it often is – push the spanner on to the nut from the right, and push up.

The diagram also shows in which direction you need to push with the spanner reaching the nut from below or above. These directions are for undoing the nut; if you want to tighten it, you push the opposite way.

Unfortunately, though, the bolt does not usually remain motionless. We need to keep it still by fitting a second spanner to the head of the bolt. The second spanner, held in your other hand, will be pushed in the *opposite direction* from the spanner used for undoing the nut.

As you push harder on the first spanner, you must push harder on the second as well. It is sometimes a toss-up whether it is the bolt or the nut which starts moving. It really doesn't matter. It can be easier to keep the nut motionless while you undo the bolt.

If the nut and bolt will not come apart, no matter how much strength is applied, try squirting on some easing oil (from motorist's shops). Wait five minutes, and try again.

Many plumbing assemblies consist of a nut which holds a pipe tightly on to a fitting, or boss, i.e. a union joint. But you simply can't see any screw-thread, so you won't know which bit to turn. See Fig. 27.

Of the two parts which look like nuts, one is permanently fixed to the pipe, and the other is the one you must undo. The one which won't move is there so you can hold the pipe steady with one spanner while you exert all your strength on the other, moving, nut.

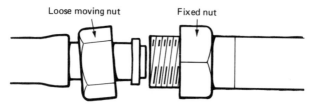

Disconnected union joint

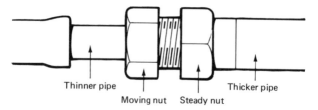

Assembled union joint

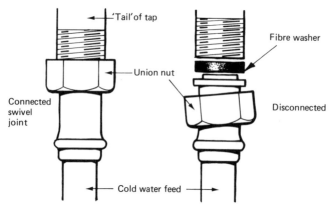

Fig. 27

If you didn't hold the pipe steady, it might move, twist, and rupture – with unfortunate results. The trouble is, it's difficult to tell them apart. Look at the drawings (left): the nut which *can* turn will have a smaller pipe coming out from its opening. Sometimes this smaller-diameter pipe will be small only for a short distance – just long enough for the moving nut to move over when it is unscrewed.

To undo it, face the back of the moving nut so you can see the smaller pipe coming out of it; hold the stationary nut with one spanner, and turn the moving nut anti-clockwise with the other spanner.

All screws move anti-clockwise to undo, and clockwise to tighten. Always use the biggest screwdriver which will fit into the slot. If you use a screwdriver blade which is thinner than the slot, you stand an excellent chance of the blade slipping out and damaging the head to such an extent that *no* blade will get the screw out.

Always, but especially when using a screwdriver with too thin a blade, push down firmly on the screwdriver handle just before – and during – the moment when you actually turn the screw.

If the screw won't come out, use a different method: hold the screwdriver as though it is a dagger. Grip the handle firmly, while pushing the blade into the slot of the screw and twisting the screwdriver anti-clockwise. Twist it just enough so it won't come out of the slot, and bang the top of the screwdriver with a hammer. Twist more and more strongly while giving harder and harder hammer taps. Eventually (I hope) the screw will begin to turn. When it does, undo it in the normal way.

9 Blocked pipes

The plumber's mate To clear a blockage in a U-trap or waste pipe, first try the plumber's mate. Place the rubber cup over the hole at the bottom of the sink through which the water normally leaves. Push down on the stick, firmly and quickly, three or four times in rapid succession.

A lot of air, dirty water and noise will come from the overflow, so you should block the overflow with an old rag held in place with one hand while you push the plumber's mate with the other. This is difficult but not impossible.

If you have no plumber's mate, you can try cupping your hand over the waste, palm down. Pump it up and down as though to push water down the hole. If you do this quickly, while you block the overflow with your other hand, you can sometimes clear slight blockages.

Spiral curtain wire If the plumber's mate doesn't clear the blockage, try wriggling a length of spiral curtain wire down the waste. If you turn it as though it were a screwdriver, while at the same time pushing down, you can often work it all the way around the U-trap. This is in fact the only method to use if you can't get at the U-trap (as, for instance, in a bath which has been boxed in).

Opening the trap If you cannot clear the blockage using a plumber's mate or a spiral curtain wire, you may have to open the trap.

First, identify the type of trap (see Section 19, Figure 67).

The U-trap usually has a plug at the lowest point of the U underneath the tube; occasionally the plug is fitted to the side of the U-tube at its lowest point.

To undo the plug which screws into the metal of the U-trap, turn it anti-clockwise as seen from below. It will be easier if you place a bar of metal (a screwdriver, or an old fork-handle or a file) between the two lugs while you do this. Occasionally you may come across another type of lug to the plug (shown in the diagram, right) which will require a pair of pliers to grip the ridge across the centre, and so to twist the plug.

Be sure to place a bucket underneath the trap before you open it. When the plug has been undone and has actually come out, all the water in the sink will rush out. If it doesn't, then the blockage is between the plug and the hole in the sink. In that case, push the screwdriver (or a straightened wire coat-hanger) up into this part of the pipe to push the blockage away.

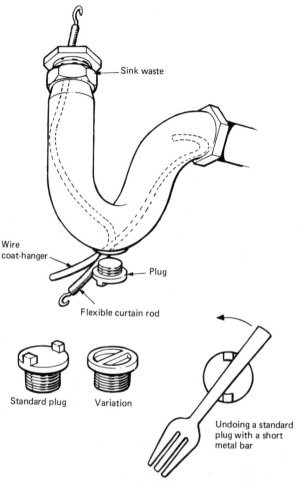

Sink waste

Wire
coat-hanger

Plug

Flexible curtain rod

Standard plug   Variation

Undoing a standard
plug with a short
metal bar

Fig. 28

More usually, though, the blockage will be in the section of pipe away from the sink. Try poking the coat-hanger or a piece of spiral curtain wire up the U-trap away from the sink, wriggle it around until it goes in all the way. Refasten the plug (clockwise) and try out the sink by running water into it. You may have to do this several times before the water runs away satisfactorily.

With bottle traps, or plastic U-traps, matters are somewhat easier. Plastic bottle traps are undone by gripping the bottom with your hand (see Figure 30) and unscrewing.

Plastic U-traps feature collar nuts which unscrew using simple hand-pressure.

For these types of wastes, clean out the blockage using a wire coat-hanger or spiral curtain wire, as above.

If all your efforts to clear the waste have failed, then you'll have to hire a plumber who'll come along with a long, man-sized version of the spiral curtain wire to push down the waste (a plumber's snake). This will bore its way to the blockage and clear it out. You can hire these snakes and do the same job yourself. Compare the cost of hiring a plumber with the cost of hiring one of these devices.

Prevention being better than cure, you should clear out U-bends and traps beneath sinks and basins regularly, not just when they are blocked; add some caustic soda or patent drain cleaner regularly to prevent the gradual build-up of matter.

Blocked w.c.s

Blocked w.c.s can only be dealt with by using a special type of plumber's mate with a metal collar. Push it down vigorously into the trap several times – but not too hard or you may break the porcelain bowl. Repeat this at three minute intervals. If you fail to clear the blockage, phone for a plumber or hire an even larger version of the metal snake used to clear waste pipes.

Blocked waste-pipes in frosty weather

When it is really cold outside, you may find that water freezes inside the pipes. This may happen even to inside pipes, if you have no central heating or if it is switched off.

Plastic U-trap

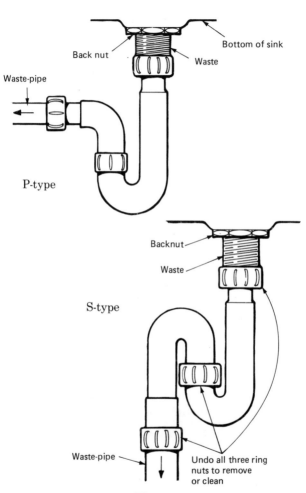

Bottom of sink

Back nut

Waste

Waste-pipe

P-type

Backnut

Waste

S-type

Waste-pipe

Undo all three ring
nuts to remove
or clean

Fig. 29

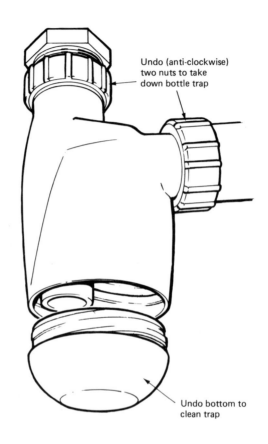

Undo (anti-clockwise) two nuts to take down bottle trap

Undo bottom to clean trap

Bottle trap

Fig. 30

Dripping taps frequently contribute greatly to this problem. Each drip, running down the waste-pipe, adds a little ice, whereas, if water goes out in one whoosh, there is much less chance of ice forming.

If you have a dripping tap, place a basin or bowl below it until you fix it. And in cold weather, don't use cold water first thing in the morning: add a kettle of boiling water to basin or sink before pulling the plug.

Keeping traps and waste-pipes clean is especially important when winter comes, since partially blocked pipes encourage the formation of ice. If, despite these precautions, the waste becomes blocked with ice, put all your kettles and large saucepans on the stove, and pour boiling water over the waste-pipe *outside* the building. This may involve using ladders or chairs, or leaning out of the window, so be very careful; rungs and sills can be slippery in that kind of weather.

Pour boiling water on to the pipe, starting at the highest point and allowing the water to drip along the pipe. Doing this two or three times will usually clear the blockage.

Blocked water pipes in frosty weather

It is rare for the air inside a house to be so cold that the water pipes freeze. This can be prevented by keeping some warmth in the house, or by lagging (wrapping almost any protective material around the pipes) all pipes where they pass through attics or cellars.

If the pipes do freeze, you can heat them using a blow-torch (don't go near any joints) or a hair-drier. A hair-drier is especially useful for heating pipes near joints and valves, which are the places at which frost blockages are most common. The water is slowed down at these points, and the tube is narrower.

Burst pipes

As water freezes inside a pipe, it expands slightly. This may not sound very serious, but the slight expansion may be just more than the pipe can take, and it will crack. You should make quite sure you know where the stop-cocks are before attempting to de-ice a pipe so that you can quickly turn your water system off before the whole house floods.

Draining the system

If you are planning to be away from your house for any length of time during very cold weather, you should drain the entire system. Switch off the water at the main stop-cock. There should be a draw-off tap next to it, to which you attach a hose to drain off all the water into the garden. If the lowest point is in the cellar, the last bit of water may have to be drained into buckets.

When all the water is drained off, open all the taps in the house, and blow mightily, as if filling a balloon, into each tap. Either it is clear, or you will have blown the water into a section which is clear.

As a child, I had a neighbour with eight children. Each evening during an exceptionally cold winter, he would station a child at each tap – to clear the entire plumbing system for the night.

10 Leaks

These may be leaking joints, leaking pipes, leaking taps, and overflows which spout water all over the pavement. If any of these occur, find your stop-cock (See Section 7) and turn off the water. Then sort out the mess.

Leaking joints

First identify the type of joint, whether it be capillary or compression in copper, a lead joint, or a combination of two different metals. (See Section 6.)

If the leaking joint is a capillary joint, you will require a blow torch, flux and perhaps a new fitting in order to fix it. If you haven't got these items, or if you're in the middle of a formal reception, try this dodge:

After the water is off, wipe the tube dry with a *hot*, dry cloth; wind electrical insulation tape round and round the

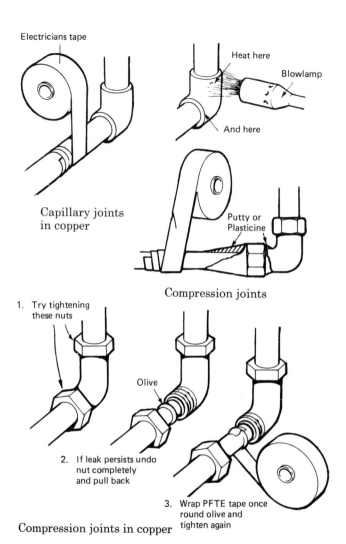

Electricians tape

Heat here

Blowlamp

Capillary joints
in copper

Putty or
Plasticine

Compression joints

1. Try tightening
these nuts

Olive

2. If leak persists undo
nut completely
and pull back

3. Wrap PFTE tape once
round olive and
Compression joints in copper tighten again

Fig. 31

tube, starting about 3 in. (7 cm) from the leaking joint. Make sure the tape laps itself about half its width. Use plenty of tape, and pull to stretch it as you go along. The tube or joint may be just a fraction of an inch away from the wall, but even so, you can pull the whole water pipe away from the wall to get behind it.

This emergency repair will stop most of the damage, but don't leave a proper repair any longer than you have to.

If a compression fitting is leaking, you won't really need to turn the water off before you fix it unless the leak is rapidly turning into a flood. Simply get out your wrench and tighten the nut (clockwise, but see Section 6 for greater detail) on the joint which is leaking. If it still leaks after you've really tightened the nut, you will have to turn the water off after all, loosen the nut all the way (anti-clockwise) until it comes off the fitting and slides along the copper tube. Pull it back a few inches, and you'll see a copper ring fitted round the copper tube where it disappears into the joint. Usually, if you pull back the copper tube, you can make it come out of the fitting far enough so that the copper ring (the olive) is fully exposed.

Wrap some PFTE tape round the pipe at the point of the olive (one complete turn will do) and push the pipe back into the socket of the joint. Do up the nut, screw it tightly clockwise with the wrench, and you should now have a water-tight joint.

Other joints which leak are the connectors between copper and/or steel pipes. For connectors to lead pipes use electrician's tape, or treat them as for lead pipes (see method below, under 'Leaking pipes'). For leaking joints in steel pipes, you will need electrical tape and Plasticine, or better still, putty.

Make sure the pipe is *really dry* all the way round; knead the putty or Plasticine until it's pliable, and then wrap it around the pipe so as to smooth out the crevices between the joint and the pipe, or between different parts of the joint.

Once a smooth profile has been made, wrap the tape around and around just as with copper tube. This is only first aid.

Leaking pipes

For leaking copper tube, treat it with electrical tape as described for leaking capillary fittings.

For leaking lead pipe, you can try electrical tape, but usually the surface of the pipe is not smooth enough for that to work. For minor holes, you might try tapping the lead around the hole with a hammer, as though you were pushing the lead towards the hole with each blow.

For long-term repairs, the section of copper or lead which is leaking must be replaced. Replacing copper tubing is something you can do yourself (see Section 6).

The replacement of lead pipe is more of a problem. When you start getting leaks, it is usually the beginning of the end; it is a sign that the whole lead pipe is rotten and requires replacement.

It is best to call in a plumber (see Section 2), ask him to fit a new lead-to-copper joint which will take a new stop-cock (see Section 6). After that, you can replace the whole of the lead pipe plumbing system yourself, with shiny new copper. The old lead pipe might even fetch a few bob from a scrap merchant.

Leaking taps

These will require a new washer, which is described in Section 11.

Overflows

When water splashes on to the pavement or into the garden from a pipe high up on the wall of your house, you need to look at the ball-cock of your w.c. or main cistern. (See Section 14.)

11 Changing a tap-washer

If you have a dripping tap, four things may happen. Firstly, if left too long, it will stain or even dissolve the enamel on your bath. Secondly, if it is a hot tap, it will waste energy.

Thirdly, if it is a cold winter, a dripping tap is the best method to ensure that your waste-pipe will be totally blocked when you get up in the morning after a hard night's freeze. Lastly, the sound of a dripping tap can be extremely irritating. So, we need to change the washer.

Tools

An adjustable spanner, an ordinary pair of pliers, adjustable pliers, big and little screwdrivers.

Materials

A washer of the right size and type. The best way to be sure of obtaining this is to take the old washer and jumper (see text below for explanation) to a good ironmonger's. This means doing the job on a week-day or Saturday morning. If you can't manage such a time, buy several, as follows:

$\frac{1}{2}$ in. (13 mm) diameter, hot and cold

$\frac{3}{4}$ in. (20 mm) diameter, hot and cold

As the washers cost only a few pence each, buy two of each type to use as spares.

Time

Actually changing the washer might take you half-an-hour the first time, and eventually about five minutes. But... first you must shut off the water supply, and that'll take time while you sort out what to turn off (see Section 7). Secondly, you may need to go to the ironmonger's for those washers while the tap is disassembled. So, the first time, allow a morning.

The job

First, shut off the water supply (see Section 7). Next, identify the type of tap, which may be one of the following types:

Pillar tap (Fig. 32) This is the oldest type, found in most houses. Begin by turning off the water supply and opening the tap, in order to drain all the water. Next, look at the part of the tap labelled 'Cover' on the drawing; you will want to undo this part. Apply an adjustable spanner at the

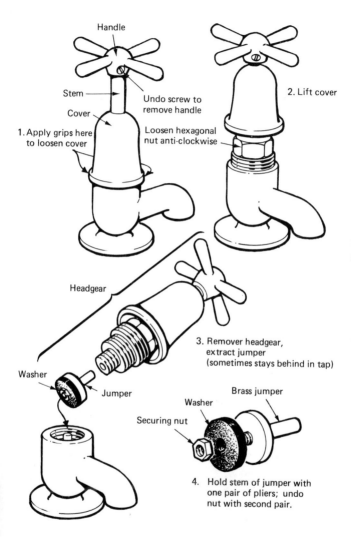

Handle

Stem

Cover

1. Apply grips here to loosen cover

Undo screw to remove handle

Loosen hexagonal nut anti-clockwise

2. Lift cover

Headgear

3. Remover headgear, extract jumper (sometimes stays behind in tap)

Washer

Jumper

Washer

Securing nut

Brass jumper

4. Hold stem of jumper with one pair of pliers; undo nut with second pair.

Fig. 32

point marked by an arrow on the drawing; tighten the spanner and turn anti-clockwise. Sometimes this part is very tight, and won't be loosened by an adjustable spanner. Wrap the cover, as it is called, in Sellotape or in cloth, and use an adjustable wrench; the wrapping is used so that you don't leave nasty scratches on the chrome cover.

When the cover is off, you can lift it till it is stopped by the tap handle, and you'll see a hexagonal nut.

Apply your adjustable spanner to this nut, and the whole of the tap will come off. Sticking out of the end of the part that has come off will be the jumper. Occasionally it gets left in the part of the tap still screwed to the sink or basin. In any case, it is just a loose piece of metal (hence it is called a jumper) holding the washer by a nut.

Hold the shaft of the jumper with a pair of pliers, and undo the nut at the end with an adjustable spanner or another pair of pliers. Now the old washer will come off; replace it with a new washer (remember that hot and cold taps need different washers) and screw the nut back on again. You can, if you wish, buy a combined washer and jumper.

The jumper gets pushed back into the hole at the bottom of the tap top, which in turn gets screwed back into the tap bottom. Tighten the tap top with a pair of adjustable pliers, and then screw back the cover, using hand pressure only. The cover is only a cover, it doesn't hold back the water.

Turn the water back on again, and the job's finished.

Modern pillar taps (Fig. 33) These can be of two types; in the first type the tap is opened all the way (after the water has been turned off) and then, as the tap handle is turned even further, the whole top comes off. In the second type, the handle comes off after you have prised a little disk let into the top of the tap (usually a different colour and marked to show H or C) with a small screwdriver, and undone the screw thus revealed with a larger screwdriver. Occasionally the little disk is stuck down, but more usually it is a snap-fit.

Once the top is off, the rest of the job is the same as for the old-fashioned pillar tap.

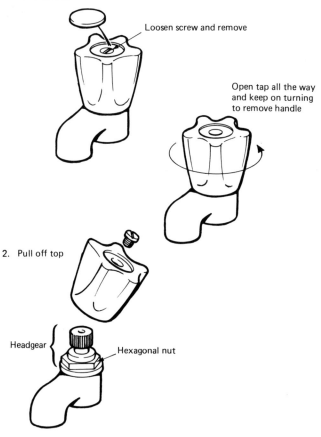

1. Undo cap with screwdriver by prising loose

Loosen screw and remove

Open tap all the way and keep on turning to remove handle

2. Pull off top

Headgear

Hexagonal nut

Fig. 33

Supataps

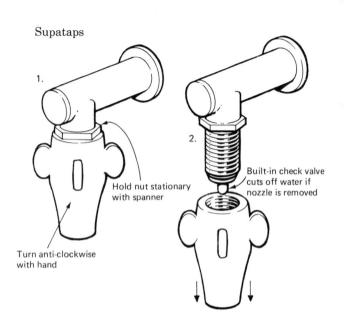

1.

Hold nut stationary with spanner

Turn anti-clockwise with hand

2.

Built-in check valve cuts off water if nozzle is removed

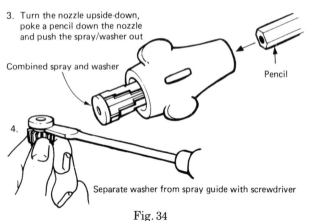

3. Turn the nozzle upside-down, poke a pencil down the nozzle and push the spray/washer out

Combined spray and washer

Pencil

4.

Separate washer from spray guide with screwdriver

Fig. 34

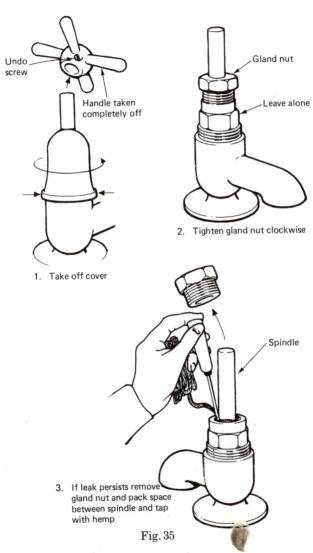

Undo screw

Handle taken completely off

1. Take off cover

Gland nut

Leave alone

2. Tighten gland nut clockwise

Spindle

3. If leak persists remove gland nut and pack space between spindle and tap with hemp

Fig. 35

Supataps (Fig. 34) With these, you don't have to turn the water off. Grasp the part shown by the arrow in Fig. 34, using an adjustable spanner, and hold it stationary while you turn the rest of the nozzle anti-clockwise, just as if you were turning on the tap. You may find that the two are so tight that they won't separate. Then you must wrap the 'ears' in cloth and use an adjustable pair of pliers as well as the adjustable spanner.

When the whole nozzle is off, push a pencil up the hole where normally the water comes out; a plastic finned flow-straightener should come out, together with the washer. The washer is a plastic combined washer and jumper; replace the whole unit, and re-assemble the tap.

Leaking gland If, after replacing the washer, the tap still leaks, it may be that water is oozing out from between the cover and the spindle. The gland nut needs tightening.

There's no need to turn the water off. First of all, you will need to undo the handle of the tap. Modern taps have a combined handle and cover, and I've already shown you how to undo these under the section on replacing washers. Old-fashioned taps have a small screw let into the side of the handle; undo this all the way, and try to pull the handle up and off. Usually it is rusted into place, and will need gentle banging from underneath to get it off.

After the handle is off, undo and take off the cover. You should now see two nuts, or occasionally one nut and a knurled nut. The lower one is the one you undo to take the whole tap off – leave it alone. Tighten the top nut or knurled nut a half-turn. If this doesn't stop the leaking, turn off the water supply and undo the gland nut (as this top nut is called) till it comes out. Push, using a screwdriver, one turn only of hemp or string (or buy the special rubber washer if you have the time) into the space between the shaft and the body of the tap, and replace the gland nut, tightening it down. Re-assemble the tap; if you have also replaced the washer, all leaking should have stopped.

12 Fitting taps

Materials

There are three major types of tap: bib-cocks are used over sinks that don't have a draining board with holes for the taps; individual taps are fitted to wash basins or to units with only cold water; mixer taps are fitted to baths and sinks.

Make sure that the horizontal distance between the tail of the tap and the spout is enough so that the water comes into the sink; make sure that you can fill a bucket *and* lift it out when checking the distance between base and spout.

You will also need special washers between the body of the tap and the retaining or backing nut; different washers are needed depending on the material of the sink or basin. Check this with the ironmonger when you buy the tap.

If you are replacing existing taps, that's all you need. For a new sink or basin, you will also need swivel fittings to connect the tail of the tap to the cold and hot water feed. I would also recommend that you fit a stop-cock at the same time since it will make life much easier in future when changing tap-washers. You will also need flux or PFTE tape, depending on whether you use capillary or compression fittings.

Tools

Basin wrench, adjustable wrench, a pair of adjustable pliers, screwdriver or steel bar (for the basin wrench), blowlamp if you're using capillary joints. You will also need woodworking tools for taking off and putting back on the sink or basin.

Time

If you can take down the basin or sink, changing the taps

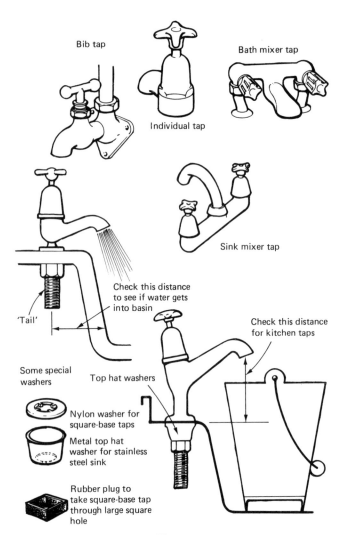

Bib tap

Individual tap

Bath mixer tap

Sink mixer tap

Check this distance to see if water gets into basin

'Tail'

Check this distance for kitchen taps

Some special washers

Top hat washers

Nylon washer for square-base taps

Metal top hat washer for stainless steel sink

Rubber plug to take square-base tap through large square hole

Fig. 36

should take you about half-an-hour per tap; add half-an-hour for switching the mains on and off (both hot and cold if necessary). Disconnecting the old taps from the mains and disengaging the waste and trap below the sink may take another hour, depending on the materials used and the age of the fitting; lastly, removing a sink top from its base or taking down a wash basin may take something like an hour and a half. I would say that at the worst you should allow six hours for the complete job, but don't be surprised if the whole thing takes only three hours.

The job

Disconnect the waste from the trap (see Section 19) and then turn off all the water supply to the taps (see Section 7). You can now disconnect the tap from the hot or cold water feed. To do this, look at Fig. 37:

The left hand drawing shows the tap connected to the water supply feed as it would look if you smashed the porcelain basin and took all the pieces away. On the right you see the same, but disassembled. To disconnect the taps, you must undo the ring nut of the swivel joint all the way. Usually this nut can be reached with a Stillson's or adjustable wrench or spanner; very occasionally you will have to resort to a basin wrench. The nut unscrews anti-clockwise, and eventually the water feed comes free from the tail of the tap. Disconnect both hot and cold taps.

Now try to take off the sink or basin. You've undone the taps, the waste and now you must undo any other retaining nuts and bolts. If it still won't come off, get someone else to try to lift it while you crouch underneath to see what's holding it down.

If it doesn't come off, as will be the case with a cast-iron bath or an old ceramic sink, then you will have to reach with a basin wrench and undo the backing nut (see diagram). I don't envy you the job, but I assure you it can be done; it merely takes time, blood and sweat.

If the unit does come off, take it down and turn it over. Generally you can undo the backing nut with an adjustable spanner or adjustable wrench; there will be no skinned knuckles and the job is easier than with the sink or

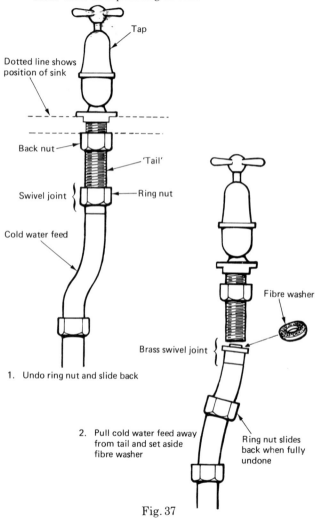

Rear view of tap sitting in sink

Tap

Dotted line shows
position of sink

Back nut

'Tail'

Swivel joint

Ring nut

Cold water feed

Fibre washer

Brass swivel joint

1. Undo ring nut and slide back

2. Pull cold water feed away
from tail and set aside
fibre washer

Ring nut slides
back when fully
undone

Fig. 37

82

whatever in place. That's why it is worth the extra trouble of disconnecting the waste and dismantling the unit.

When putting pressure on the spanner undoing the backing nut, apply opposite (clockwise) pressure with your other hand, using the spout of the tap as the lever arm of a second spanner. This is to prevent the whole tap turning, which may damage or even crack a basin or sink.

When the backing nut is completely screwed off the tail, you can withdraw the tap. Use a hammer and an old screwdriver to clear the mess of old putty and paint that may be left behind. When everything is clean, you can start with the new tap.

The most complicated part of fitting a new tap is choosing (and using) the right washers. Taps are made with either a shoulder between the tap base and the tail, or without a shoulder. The shoulder may be deep or shallow, square or round. The general idea is that the shoulder sits into a square opening let into the basin so that the tap doesn't turn round after it has been fitted. This works fine if the tap fits into a thick ceramic wash basin; if the tap is to be fitted into a thin plastic, enamel or stainless steel unit, then the shoulder is generally too thick and will prevent the backing unit from gripping the tap unit tightly. So a special washer called a top hat is needed – ask for it when buying a tap to be fitted to a thin unit.

Other taps have no shoulder at all; if they are fitted without further ado into a large square hole they wouldn't stay in place unless you tightened the backing nut so tightly that there would be a danger of cracking the porcelain, the plastic or the enamel. So a spacing washer has to be used, the thickness of which depends on the material of the sanitary unit.

To fit the tap, start with the rubber or fibre washer that comes with the tap if there is *no* shoulder. If there is a shoulder start with a ring of putty applied round the tail of the tap where it will sit on the bath or whatever. Next, push on the top hat or the spacing washer, where required, and push the tail of the tap through the hole. Make sure the base of the tap sits on the top surface of the basin, and that the top hat or spacing washer holds the tail snugly in place.

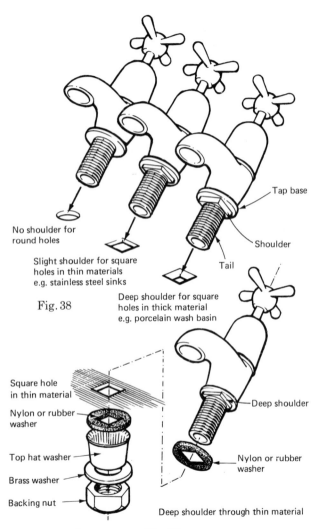

No shoulder for round holes

Slight shoulder for square holes in thin materials e.g. stainless steel sinks

Fig. 38

Deep shoulder for square holes in thick material e.g. porcelain wash basin

Tap base

Shoulder

Tail

Square hole in thin material

Nylon or rubber washer

Top hat washer

Brass washer

Backing nut

Deep shoulder

Nylon or rubber washer

Deep shoulder through thin material

Fig. 39

Now screw on, from below, the backing nut and tighten it till the tap is held firmly. There is no need to over-tighten the backing nut if the proper washers are used; in fact, too much force will cause the porcelain or enamel to crack.

If you have to renew the taps with the bath or sink left in place on its stand, you may find it easier to do all this with a helper. If not, now is the time to put the sink back on to its base, and bolt it down as you found it.

Reconnect the waste to the trap, and reconnect the tails of the taps to the water feeds. Remember the fibre or rubber O-ring that sits inside the coupling nut; replace it if missing or worn.

Turn the mains water on again; run the new taps to check the waste-pipe for leaks, and then switch them off to check the new taps and the connections for leaks. Tighten any nuts on the swivel coupling if required; you may even have to turn the water off, apply PFTE tape on the tail of the tap and screw the nut back on again before achieving a water-tight fitting.

If the tails of your new taps are too short, or if you are fitting the sink or basin for the first time, you will have to turn to Section 13.

13 Connecting up

Whether you are installing a main or w.c. cistern, a bath or sink or a wash basin, there always will come the moment when you have fixed the appliance to the wall and you need to connect up to the main water system. Connecting up requires a special procedure, and a special fitting.

Always fit and fix your appliance first. Build your sink base, screw on your brackets, build your cistern platform first, then fit the appliance into place and fix it by hand-tightening only all bolts and nuts. Be prepared to loosen and take down the appliance at least once before the final fixing.

All these appliances will be fitted with either a tap or a

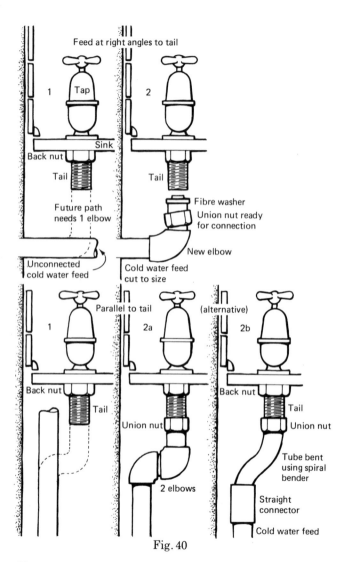

Fig. 40

Fitting a stop-cock to save turning off the whole system when you change a washer

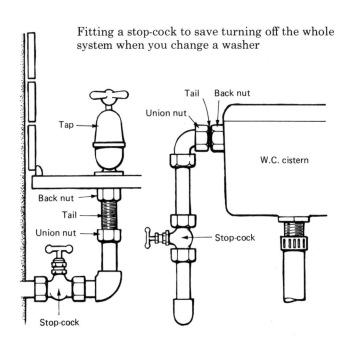

Tap

Tail Back nut

Union nut

W.C. cistern

Back nut

Tail

Union nut

Stop-cock

Stop-cock

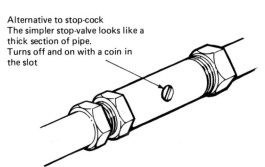

Alternative to stop-cock
The simpler stop-valve looks like a thick section of pipe.
Turns off and on with a coin in the slot

Fig. 41

ball-cock (see Sections 12 and 14) which must be fitted first. After fitting, a tail will be left sticking out from the appliance.

Having fitted the appliance, and having fitted the tap or ball-cock, turn your attention to the water feed. Run this feed towards the appliance; if you can, have the feed running at right-angles *past* the tail as in the picture on the left in Fig. 40. If the feed runs parallel with the tail, more work is usually required to connect the tail and the feed.

The final connection between feed and tail is made using a special connector called a swivel joint. Fit this connector first, but only hand-tighten the nut. Now cut the feed, and use elbows and small lengths of pipe to connect the feed to the final connector. Generally speaking, the use of two elbows will allow you to connect any two points. If you have a tube bender, you can bend the tube; this is especially handy if you have the feed running along or even in the wall and the tail of the tap is about 3 inches from the wall.

Once you have cut all your piping in a dry run (without having soldered capillary joints, or tightened compression joints), you undo the connector fitting and pull it away from the tail. Now you make all your joints in earnest, checking all the while that the pipes are in the right position in relation to each other, finishing up with the coupling fitting.

If the water feed is buried in the wall or otherwise fixed, you may now have to lift or move the appliance slightly so that the tail of the tap or cock can be engaged by the nut of the special connector. More usually, there is a slight 'give' in the feed, so that it can be bent the necessary $\frac{1}{4}$ in. (8 mm).

If, after soldering or tightening all the joints of the feed, the special connector doesn't quite reach the tail, you may have to loosen the joints slightly, re-position the tubes, and try again. Compression fittings can be tightened and loosened (without being actually disconnected) merely using a spanner. Capillary fittings are re-positioned by playing the flame of a blowlamp over the joint; hold the fitting with a pair of adjustable grips and twist it in the direction you want it to move, which it will do as soon as the solder is soft. After re-positioning, tighten up the connector fitting.

Incidentally, if you have a little extra cash, always fit a stop-cock to the feed just before the special connector.

14 Ball-cocks

When you explored your plumbing system and came to either the main cistern, or any of the cisterns above your w.c., you must have noticed that the cisterns all had two pipes stopping off fairly near the top. One of these is the water supply pipe; its supply of water to the cistern is regulated by a special valve called a ball-cock. The ball-cock lets more water in after you have flushed the w.c. or drawn off water for the bath, and stops off the water supply after the tank is full. If it doesn't stop it off properly, water will come out of the other pipe, the overflow.

You'll notice that something is wrong with the ball-cock when water starts dripping, or sometimes even pouring, out of a pipe on the outside of your house.

Open up the top of your cistern; w.c. cisterns sometimes have two screws in the centre of the two sides, and sometimes four screws, one in each corner. When you look inside, you will see a large round ball floating on the surface. If it is the original fitting it will be of brass, often very much rusted. Nowadays this ball-float, as it is called, will be of plastic. The ball will be about 4 in. (100 mm) in diameter. If you push this down into the water, more water will enter the tank, and if you pull it up, the water will stop. That's exactly how the valve works; when you flush the w.c. the water sinks, the ball sinks with it and more water comes into the cistern. As the water level rises, the ball rises, and gradually shuts off the water supply.

There are three causes of faulty ball-cocks: the ball leaks, the float arm is too high, or the valve is worn out.

Ball leaks

To check for leaks in the ball, take hold of it firmly and unscrew it (anti-clockwise) from the ball-float arm. When it is off, shake it to hear if there is any water inside. If there is, replace it with a modern plastic ball.

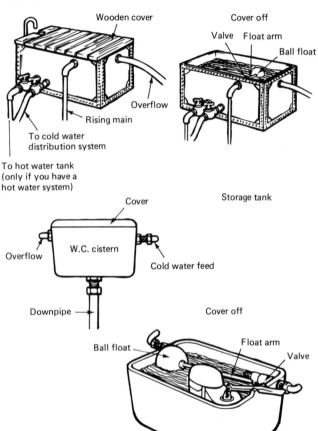

Overflow from hot cylinder
(only if you have a hot water system)

Wooden cover

Cover off

Valve Float arm

Ball float

Overflow

Rising main

To cold water
distribution system

To hot water tank
(only if you have a
hot water system)

Cover

Storage tank

Overflow

W.C. cistern

Cold water feed

Downpipe

Cover off

Ball float

Float arm

Valve

Fig. 42

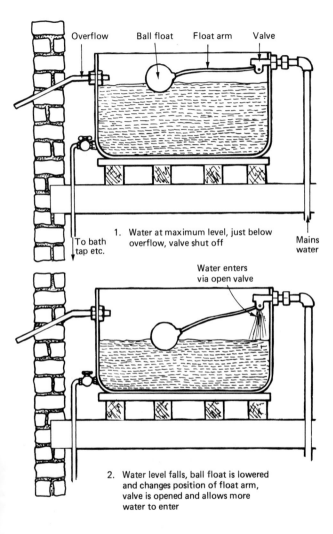

1. Water at maximum level, just below overflow, valve shut off

2. Water level falls, ball float is lowered and changes position of float arm, valve is opened and allows more water to enter

Fig. 43

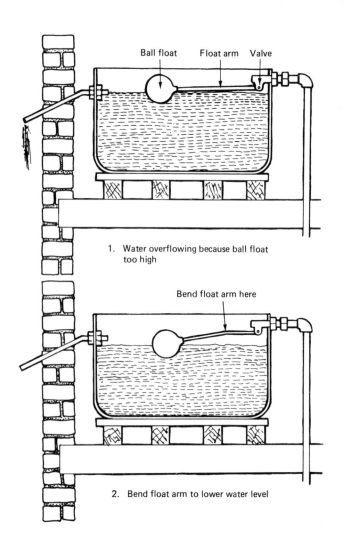

1. Water overflowing because ball float too high

2. Bend float arm to lower water level

Fig. 44

Float arm is too high

If you're dealing with a w.c. cistern, simply flush the w.c. If you're dealing with a main cistern, open up all the cold water taps in your house till the cistern is half-empty. In both cases, pull the float arm up as high as it will go, and water should stop coming in. If any water, even a slow drip, continues coming in, you will need to deal with the valve itself. If the water stops coming in completely as you raise the float arm, then you will need to bend this float arm a little so that the ball sits lower. The float arm is made of soft metal, and will bend fairly easily. Let the cistern fill again, and check that the new water level is below the opening to the overflow.

Defective valve

Usually it is only the washer that will need replacing. Start by turning off your water system (see Section 7). Have a good look at the valve, which is the brass fitting at the side of the tank at the end of the float arm. There are three types: the Portsmouth, the Croydon and the Diaphragm.

The Portsmouth Follow the float arm, and you will see that it hinges on a split pin, which is rather like a giant hairpin pushed through the hole, with its ends bent out to stop it coming out of the hole.

To take out the split pin, use a screwdriver and a pair of electrical pliers to push the two ends together, as in Fig. 46, and then pull the pin out altogether. Now undo the knurled knob (anti-clockwise) and lay it to one side. With a screwdriver poke out the piston inside the valve, sliding the screwdriver along the slot underneath the valve (Fig. 47). The piston consists of two parts which are screwed together to retain the washer. Slide the blade of a screwdriver into the slot in the side of the longer part, and grip the smaller part with a pair of pliers. Unscrew the two parts to reveal the washer. Try not to mark the valve with your pliers. Replace the valve, and re-assemble, greasing the piston with Vaseline before pushing it back into the valve body. Make sure that the slot in the piston sits exactly over

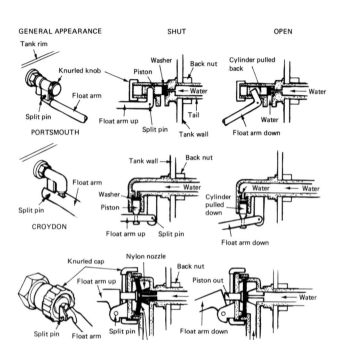

Fig. 45

the slot in the valve body, otherwise you won't get the float arm to sit properly. Push the split pin back in again, and separate the two ends. Turn the mains water on, and check that everything works properly.

The Croydon The Croydon differs from the Portsmouth in that there is no knurled knob to undo. Instead, after you have removed the split pin, the float arm will come free, and take with it the piston. Leave the piston on the float arm, and with a pair of electrical pliers, unscrew the smaller section of the piston, just as before. Replace the washer,

Removing a split pin

Section

  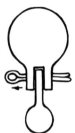

1. Lever the two bent
   ends with a screwdriver

2. Push the ends
   together with pliers

3. Push the pin out,
   or pull with pliers

 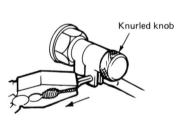

Knurled knob

Fig. 46

Removing the piston

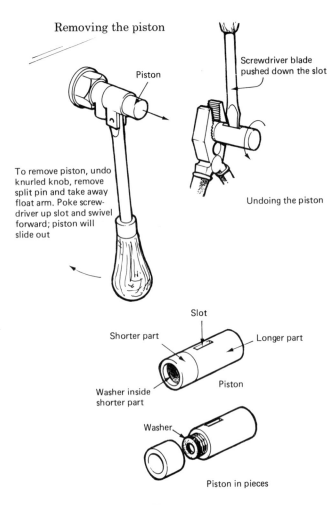

Piston

Screwdriver blade
pushed down the slot

To remove piston, undo
knurled knob, remove
split pin and take away
float arm. Poke screw-
driver up slot and swivel
forward; piston will
slide out

Undoing the piston

Slot

Shorter part

Longer part

Washer inside
shorter part

Piston

Washer

Piston in pieces

Fig. 47

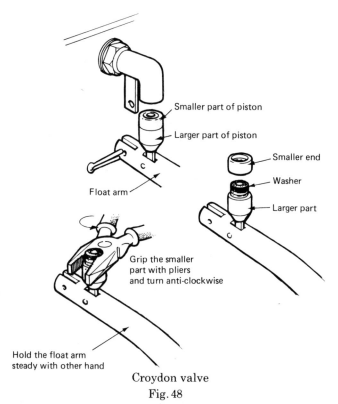

Smaller part of piston

Larger part of piston

Float arm

Smaller end

Washer

Larger part

Grip the smaller
part with pliers
and turn anti-clockwise

Hold the float arm
steady with other hand

Croydon valve
Fig. 48

grease the piston, and re-assemble. This time you will have
no problem aligning piston and float arm.

The Diaphragm Remove the split pin as before. The
knurled cap is then unscrewed using hand-pressure or
gentle application of a pair of adjustable pliers. The cover
will now come off, and reveal the diaphragm, which you
simply take out of the opening and replace. Re-assemble all
the parts, and remember to sit the cover back on to its
seating in such a way that the hinge to the float arm will be
exactly at the lowest point.

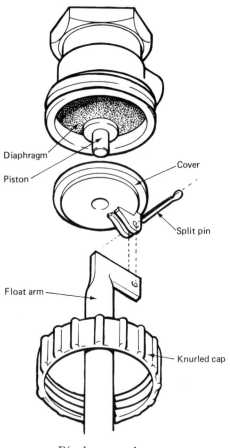

Diaphragm — Piston — Cover — Split pin — Float arm — Knurled cap

Diaphragm valve

Fig. 49

If, after bending the float arm, renewing the ball float, and replacing the washer, the valve still doesn't work properly, you will have to replace it entirely (Section 15).

15 Cold-water storage tank

Every house should be fitted with a cold water storage tank; this will keep a supply of water inside the house for w.c. and washing facilities if the mains water is cut off (e.g. for street repairs). Many old houses don't have one, and other old houses have an old one which may leak.

The tank should be fitted as high as possible, i.e. in the attic of your house. Usually there is a hatch giving access to the roof; measure the size of this hatch carefully to determine the largest cistern that will pass through. It is useful to know that some types of plastic cistern can bend and fold to some extent. The old cistern will probably be made from galvanised steel and may have been put in before the roof was finished; you won't get it out in one piece.

The new tank, made from a plastic such as polythene, should hold at least 50 gallons (220 litres), and preferably 80–100 gallons (350–440 litres). You will need to form at least three holes in the tank, using an electric drill with a hole-saw attachment. It is best to cut these holes before the tank is put into the attic, since most electric drills don't have long enough leads to reach that far.

If you're replacing an existing tank, try to buy a tank of the same shape as the old one; you will then only need to drill holes in the same position as the holes in the old one. For a new tank, drill as follows:

> One hole, 2–3 in. (50–80 mm) down from the rim of the tank to take the inlet valve or ball-cock (see Section 14). The diameter of the hole depends on the diameter of the 'tail' of the valve.
> A second hole to take the overflow, about 5 in. (125 mm) from the rim of the tank, the diameter of which depends on the overflow pipe fitting.
> A third hole, about 2 in. (50 mm) up from the bottom of the tank to take the water distribution pipe supplying

bath, wash basins and w.c. cisterns.

There may be a fourth hole for water supplying the hot water system if you have one already.

When buying the tank, you will also need to buy the various fittings, unless you are merely replacing an existing tank and retaining the existing fittings. You will need:

An inlet valve, with ball arm and float; this valve is fitted on to the end of your rising main. If you're extending the rising main to lead to the attic, use $\frac{1}{2}$ in. (15 mm) diameter copper tube, and buy the valve to fit.

$\frac{3}{4}$ in. diameter plastic overflow pipe with a fitting to secure it to the hole in the cistern; there should be enough pipe to lead to an outside wall.

$\frac{3}{4}$ in. diameter (22 mm) outlet fitting to take the $\frac{3}{4}$ in. diameter (22 mm) distribution pipe.

$\frac{3}{4}$ in. diameter (22 mm) stop-cock, to be fitted immediately adjacent to the place where the distribution pipe leaves the cistern.

Similar outlet fitting and stop-cock if you have a second distribution pipe for your hot water system.

After you have cut the holes, prepare a platform for the tank to sit on. The tank, complete with 100 gallons (440 litres), weighs about half a ton; it needs firm support. If you are using a new polythene tank, it will also need a very smooth base; use a 19 mm thick piece of marine quality plywood or blockboard laid on 2 in. × 2 in. (50 × 50 mm) softwood bearers about 6 in. (150 mm) apart. Obviously, if you have an existing tank it will already have bearers, but it is usually better to have a new baseboard. For a new tank, look for a space not too far from an outside wall, and it is also best if the tank sits over a partition or wall in one of the rooms below (remember that half-ton weight).

If you are replacing an old tank, first turn off the water at the mains, and then drain all the water in the tank by opening one or more taps below. Undo the back nut holding the inlet valve to the water tank, and disconnect the rising main water pipe by unscrewing the nut holding the union

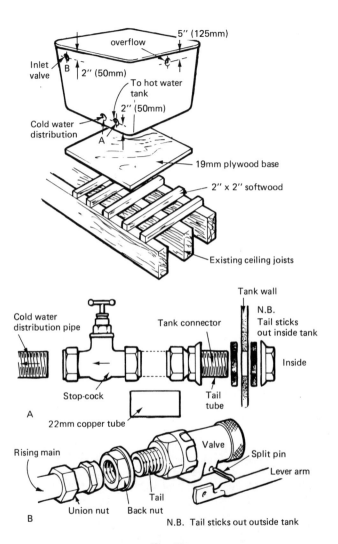

5″ (125mm)

overflow

Inlet valve

B

2″ (50mm)

To hot water tank

2″ (50mm)

Cold water distribution

A

19mm plywood base

2″ x 2″ softwood

Existing ceiling joists

Tank wall

N.B. Tail sticks out inside tank

Cold water distribution pipe

Tank connector

Inside

Stop-cock

Tail tube

A

22mm copper tube

Valve

Rising main

Split pin

Lever arm

Union nut

Back nut

Tail

B

N.B. Tail sticks out outside tank

Fig. 50

101

joint (for a lead pipe) or compression joint (for a copper tube). Repeat the procedure for the overflow and the water distribution pipe. The tank will now lift off. Put it to one side, fit the new plywood base and put on the new tank. Re-assemble the three fittings, reconnect the various pipes, and turn on the main. Check for leaks. The old tank can be left where it is if you don't need the space; if you do need the space, cut it up with a large hacksaw and judicious blows of a club-hammer.

If you are fitting a new tank, first prepare the base. Lift the new tank on to it, and then fit the fittings. Each fitting has a main body and a tail. Round the tail will be screwed a back nut, which must first be removed using hand pressure only, as well as a pair of washers. Leave on one washer, push the tail through the hole, fit the second washer, and then screw on the back nut, and tighten it with a wrench. Repeat for all three (or sometimes four) fittings. Fit the inlet valve with the ball-float. Now you are ready to install the pipework.

First, extend your rising main using $\frac{1}{2}$ in. (15 mm) diameter copper tube. If your existing rising main is in lead, I would advise getting a plumber in to replace the main stop-cock (or install one if you haven't one already) with a 10 in. (250 mm) copper tail with one T-junction (see Section 6), and take a new copper rising main from this point all the way up to the attic. If you can't afford this, or if you already have a copper rising main, simply take off your supply to the cistern from the nearest convenient point. Bring the copper tube to the inlet valve, and make the last connection using the compression joint that forms part of the tail of the inlet valve.

Next, provide a path for your overflow. The overflow pipe should lead to an outside wall of the house, which usually means angling the pipe down to the space between the joists and along till it comes to the narrow junction between roof and attic floor. With a hammer and a long cold chisel, make an opening as low as you can – try to make the opening *below* the gutter.

Fit the overflow fitting into the appropriate hole in your cistern – the arrangement for washers and back nut is the same as for the inlet valve, except that all the fittings are plastic, and therefore need hand-tightening only. The rest of the overflow pipe is either made using push fittings or fittings using a solvent glue. Leave the pipe sticking out of the wall by about 4 in. (100 mm).

Lastly, fit the water distribution fitting(s) – the general arrangement is identical to the inlet valve as regards washers, tail and back nut. This time, the tail is connected directly to a stop-cock, and the water distribution pipe(s) connects to the stop-cock outlet. The side of the stop-cock is marked with an arrow – this shows the way the water should flow when the stop-cock is fitted and turned on.

When all is fitted up, turn on the main stop-cock and check for leaks. Also check that the ball-float of the inlet valve stops off the water at the right level. If the water level continues to rise after the water level reaches the hole of the overflow, bend the arm of the ball float a little so that the ball is lowered (see Section 14).

16 Repairing w.c. cisterns

Cisterns are the tanks of water sitting on the wall above your lavatory. There are two main types; the old-fashioned bell-type cistern, usually made of cast iron and fixed at a high level above your w.c., and the newer low-level plastic cistern.

Bell-type cisterns

This type works through the weight of a cast-iron bell pushing down some of the water in the cistern after you let go of the chain. The water inside the bell has nowhere else to go but *up* into the narrower section of the bell, and so over the top of the down-tube. Once some of the water has gone into this down-tube, it pulls the rest along by siphonic action, till the water level in the cistern is so low that air can get into the bell from underneath and so break the siphon. The cistern stops flushing, and begins to refill with water coming in via the ball-cock (see Section 14).

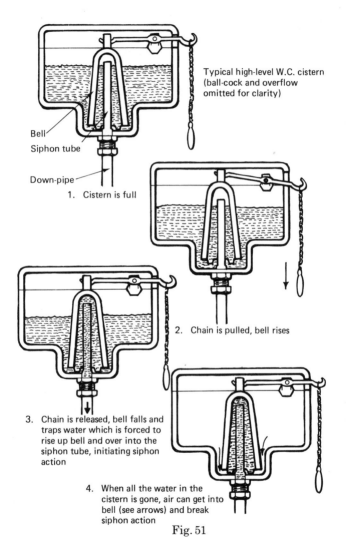

Typical high-level W.C. cistern (ball-cock and overflow omitted for clarity)

Bell

Siphon tube

Down-pipe

1. Cistern is full

2. Chain is pulled, bell rises

3. Chain is released, bell falls and traps water which is forced to rise up bell and over into the siphon tube, initiating siphon action

4. When all the water in the cistern is gone, air can get into bell (see arrows) and break siphon action

Fig. 51

If it doesn't flush properly, and you have to pull the chain several times, it usually means that the bell isn't coming down quickly enough. The parts may be corroded or grit may have got into the mechanism. I'm afraid the only remedy is to take the cover off, disassemble all the parts (see Section 8) and clean them as well as you can. Usually the cover just lifts off, occasionally it is held into place with either two screws in the centre of each side, or four screws, one at each corner. Remove rust, scale and dirt with a stiff wire brush. Re-assemble, grease moving parts, and hope that your work is sufficient. Otherwise think of replacing the whole cistern, which will take little more work than disassembly, cleaning and re-assembly.

If the water goes on flowing into the w.c. bowl after the flush is supposed to have finished, this indicates that the bottom of the cistern is filled with rubbish such as rust, grit or even bits of plaster. Then the bell simply can't drop low enough. Officially you're supposed 'simply to clean out the cistern' but it can prove to be a hell of a job with a high-level cistern and a low ceiling. Again, think of replacement.

Low-level cisterns

Modern low-level plastic cisterns go wrong much more often, frequently every two or three years, but they are somewhat easier to deal with. It is, however, difficult to describe the disassembly of one in words; trying it out in practice may make things much clearer.

Take off the top of the cistern, and look inside. You'll see the ball-float attached to the end of a long thin metal arm. Find a piece of wood and some string, and tie the ball-float arm to the piece of wood, which should sit on top of the cistern as a sort of bridge. The general idea is to keep the ball-float as high as it can possibly go, since this will stop further water coming into the cistern. Next flush the cistern; if it won't flush it will have to be emptied with a cup or a siphon tube.

Now look underneath the cistern and identify the ring nut holding the flush pipe to the siphon tail. Undo this nut using hand pressure, or perhaps with slight assistance from

Low-level cistern

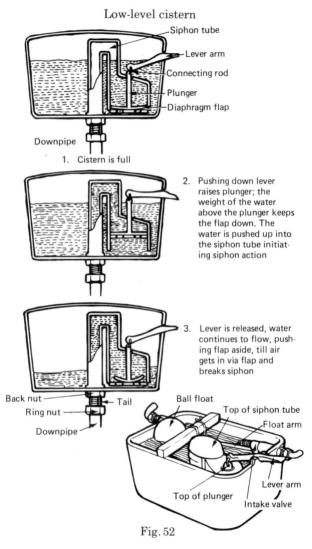

Siphon tube
Lever arm
Connecting rod
Plunger
Diaphragm flap

Downpipe

1. Cistern is full

2. Pushing down lever raises plunger; the weight of the water above the plunger keeps the flap down. The water is pushed up into the siphon tube initiating siphon action

3. Lever is released, water continues to flow, pushing flap aside, till air gets in via flap and breaks siphon

Back nut
Ring nut
Downpipe
Tail

Ball float
Top of siphon tube
Float arm
Top of plunger
Lever arm
Intake valve

Fig. 52

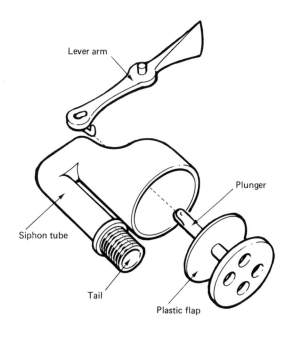

Siphon tube removed from w.c. cistern

Fig. 53

a pair of adjustable pliers. When it is all undone, the remainder of the water in the cistern will come out.

The tail of the siphon tube will be held to the bottom of the cistern with a back nut, which you must undo next, again using hand pressure or slight pressure from a pair of adjustable pliers. When the back nut is completely off, stand up again and look into the cistern. Identify the arm attached to the lever you push when you want to flush; attached to this arm will be a metal rod disappearing into the siphon tube. Detach this metal arm, and the whole siphon tube will lift out of the cistern with a bit of wriggling.

Turn the siphon upside down and you will see two openings. The larger one will contain a plunger which slides up and down. Remove this plunger, and on the far side of the plunger (on top if the siphon were in its normal working position) will be a plastic flap. It is this flap which will need replacing – you can buy a new one at an ironmonger's. Take the plunger with you to check the actual size since diameters can vary.

Reassemble all the parts and clean them as you go. When all the parts are together (don't forget washers and the like) undo the string holding the float arm up, and let the cistern fill. Flush once to check that all systems are go, and put the top back on.

If you have any problems with regard to the cistern not filling up, or filling up so much that water overflows, check Section 14 on ball-cocks.

17 Replacing a w.c. cistern

Replacing a w.c. cistern is very similar to replacing a main water storage cistern, except that the holes are already pre-cut and the water distribution pipe (called the downpipe) is much thicker, and is made of plastic.

First turn off the water supply and flush the w.c. Using a cup, or a rubber tube as a siphon, get out as much of the water as possible. Disconnect the union joint (if the supply pipe is copper); disconnect the overflow pipe.

With a large wrench undo the union joint between the downpipe and the tail of the fitting to the cistern; if this downpipe is lead, the chances are that you'll wreck this joint, but don't worry, you'll need to renew the downpipe at the same time. The cistern should now lift off its brackets.

If the downpipe is lead, cut through the lead about 3 in. or so (75 mm) from the point at which it enters the porcelain of the w.c. bowl, using a hacksaw. Remove the main part, and gently (for fear of breaking the porcelain bowl) wriggle the stump out of the w.c. bowl. The stump will be held in with caulking, which should be cleaned out from the spigot.

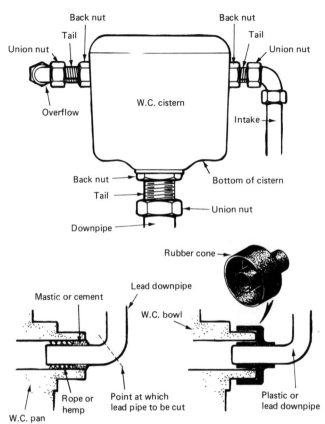

Downpipe caulked to w.c. pan   Downpipe fixed using rubber cone

Fig. 54

You can replace a high-level bell-type cistern with a modern plastic one, in which case you merely reconnect the inlet valve (this should be a new one) to the water supply pipe, disconnect the old overflow fitting and fit it to the new tank so that the new cistern can connect with the old

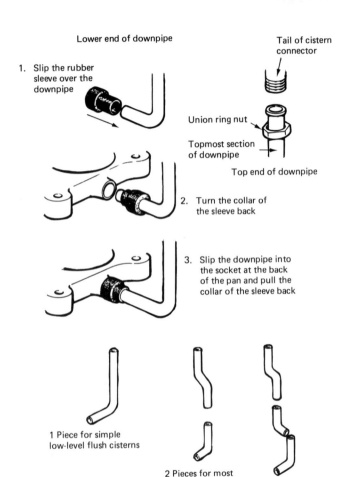

Lower end of downpipe

Tail of cistern
connector

1. Slip the rubber
   sleeve over the
   downpipe

Union ring nut

Topmost section
of downpipe

Top end of downpipe

2. Turn the collar of
   the sleeve back

3. Slip the downpipe into
   the socket at the back
   of the pan and pull the
   collar of the sleeve back

1 Piece for simple
low-level flush cisterns

2 Pieces for most
high-level cisterns

3 Pieces for offset
w.c. cisterns

Fig. 55

overflow pipe. The new downpipe will be described in the section below on a new low-level cistern.

If you're fitting a new low-level cistern (see my remarks on the difficulty of cleaning and repairing high-level cisterns), you must first mark out, in outline, the point at which the top of any future brackets will support it. Drill and plug the holes for the screws holding the brackets to the wall, and then fit your new cistern. Extend or adapt your cold water feed in $\frac{1}{2}$ in. (15 mm) diameter copper tube. Fit the water inlet valve as before, and connect the water feed pipe using a compression fitting on the end of the inlet valve's tail.

You will have to form a new hole in the wall (using hammer and cold chisel) to take the overflow pipe. Buy $\frac{1}{2}$ in. (15 mm) plastic overflow pipe and fittings and connect them to the fitting which you'll secure to the cistern, tightening the back nut using hand-pressure only.

Lastly, the downpipe. Downpipes come in two or three pieces, depending on the complexities of getting from the cistern outlet to the w.c. inlet. Cut the various tubes to fit, and assemble dry (i.e. without using glue) to check; always cut the tubes (using a hacksaw or a tenon saw) a little too long, since it is very easy to cut a little more tube off. The topmost tube connects to the tail of the cistern outlet using a large union joint with a ring nut which is tightened using hand-pressure only. The downpipe fits into the back of the w.c. bowl in the spigot, and is held in place with a special rubber sleeve. The procedure is to wriggle the rubber sleeve about 3 or 4 in. (75 or 100 mm) along the downpipe and draw back the outer collar; the downpipe is now pushed into the spigot, and the collar pushed back over the outside of the spigot extension

18 Replacing or fitting a w.c. bowl

Materials

You will obviously need to buy a porcelain w.c. bowl. They come in all colours and shapes, but all that you need to worry about is whether they have a P- or an S-trap.

Look at your existing w.c. bowl, and buy a similar one. If you're fitting a new w.c. for the first time, then you'll need a P-trap for connection to a soil-stack, and an S-trap for direct connection to the man-hole. The actual adaptation of a soil-stack or man-hole to take a new w.c. is not beyond the skill of an amateur, but does go beyond the scope of this book. Get a plumber to do this part.

You must decide whether you want to re-use the old w.c. seat or buy a new one. Plastic ones are cheaper, but tend to warp or crack; new wooden ones are very expensive but beautiful. If you have an existing wooden one, consider stripping off old varnish or paint and revarnishing it.

If you have a plastic downpipe between cistern and bowl no further parts are required. If you have a lead downpipe, think of replacing at least the downpipe (see Section 17 for details).

Lastly, to connect the w.c. bowl to a cast-iron or plastic soil-stack, you will need a special plastic connector piece. An existing toilet connected to a plastic soil-stack will already have such a connector, but one connected to a cast-iron stack will need a new one. In either case, if you are replacing the cistern at the same time, you will need a special extension connector, which allows the bowl to sit 3 in. (75 mm) or so further away.

Tools

You'll need a hammer and either a cold chisel or a sturdy

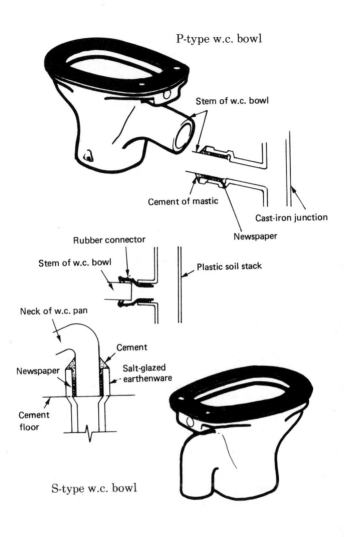

P-type w.c. bowl

Stem of w.c. bowl

Cement of mastic

Cast-iron junction

Newspaper

Rubber connector

Stem of w.c. bowl

Plastic soil stack

Neck of w.c. pan

Cement

Newspaper

Salt-glazed earthenware

Cement floor

S-type w.c. bowl

Fig. 56

screwdriver to clear out the caulking at the points where the downpipe from the cistern and the socket of the soil-stack connect to the porcelain bowl. You'll need a screwdriver to remove the screws holding the bowl to the floor. Lastly you'll need a pair of pliers to undo the nuts holding the seat to the bowl.

Time

Assuming that you have a downpipe from the cistern with a quickfit rubber connector, you should allow about an hour to remove the pan, half-an-hour for removing the old seat and fitting it to the new bowl, another half-hour for connecting the new bowl to the soil-stack and downpipe and a final half-hour for fixing the bowl to the floor. Add a further half-hour if you have to connect to a salt-glazed earthenware pipe. Altogether $2\frac{1}{2}$ to 3 hours.

The job

Unscrew the long brass screws at the foot of the bowl if it is sitting on a timber floor – w.c. bowls on concrete floors are usually cemented into salt-glazed earthenware pipes which hold them in place. Next, disconnect the downpipe (see Section 14 for details) where it enters the bowl.

You must now disconnect the bowl from the soil-stack or pipe. If there seems to be a rubber or plastic sleeve between the tail of a P-trap type w.c. bowl and the soil-stack branch, peel it back as you would to remove a rubber washing-up glove. The bowl should now come free and can be removed.

If the tail of the P-trap enters a cast-iron branch, try moving the w.c. bowl from side to side and easing it out of the socket. Sometimes the tail of the P-trap is held in the socket by a rubber cone, in which case the bowl will come out altogether if you gently but firmly pull it away from the connector.

If the bowl cannot be moved at all, then probably it is held in place by caulking topped with a quickset cement. Start by emptying the water from the bowl, using a cup or plastic beaker. You must now smash the bowl with the hammer, starting at the junction of the tail of the bowl, and the bowl itself. When you have smashed enough to remove the main bowl, and have cleared the mess of shards and water, start chipping away the rest of the tail using small taps rather than mighty blows. Eventually, when all the shards are cleared, you can clear the cast-iron socket of porcelain and cement pieces, as well as bits of old rope and newspapers.

With S-traps which connect directly to salt-glazed earthenware pipes, a similar procedure is followed, except that even greater care must be taken when you're breaking up the last bit of the w.c. If you do manage to break the salt-glazed earthenware pipe, keep all the pieces and clean them very thoroughly; by using a two-part adhesive it is possible to glue them together again.

You are now ready to put in the new pan. Knock up a small amount of cement/sand mixture (this is sold ready-mixed) with enough water to make a smooth thick mess. Make a little mound on the floor where you're going to put down the bowl, a mound about 5 in. (125 mm) diameter and 1½ in. (40 mm) high. Put your bowl on top of this mound, and wriggle it around till the bowl is sitting firmly. Check with a spirit level that the pan is level (left-to-right). You are now ready to roll back the connector sleeve if connecting to a plastic soil-stack; if you are connecting to a cast-iron soil-stack you will use either a plastic sleeve or a rubber cone.

If you are connecting to salt-glazed earthenware, stuff rolled-up newspaper or old rope in the space between the tail of the w.c. trap and the rim of the earthenware pipe, and top it up with some of the cement you have knocked up. Leave this connection to harden for two or three days, and be careful to avoid moving the bowl for the first month.

Lastly, connect the downpipe to the new bowl (see Section 14 if you have to replace the downpipe). Screw the bowl to the floor if the floor is a timber one, and clear off any surplus cement/sand mixture that shows around the foot of the bowl.

If you haven't done so already, remove the old seat; it is held in place by long bolts, with nuts clamping these down hidden underneath the old bowl, at the back. Watch the sequence of nuts and washers as you undo them, and put them on the new bowl in the same order. If you have bought a new seat still undo the old ones in order to see how to put the new ones on.

19 Waste-pipes

If you look under your sink or wash basin, you'll see strange shapes made up from pipes of various sizes twisted into bends; these pipes disappear into the floor, wall or duct and are generally about $1\frac{1}{2}$ in. (40 mm) diameter. There is a similar arrangement under the bath, although generally you can't see it because the bath is boxed in.

The whole arrangement for getting water from the sink into the main downpipe is made up of the parts shown in Fig. 57.

When you are deciding where to place a new sanitary fitting, bear in mind that the waste-pipe will be the most important component of that decision. The rule is that the waste-pipe must always slope *down* (it may never go up, even for a little way) and that it may not be more than 5 foot 6 inches (1600 mm) from trap to waste outlet (6 foot 6 inches or 1900 mm for baths). If you make this distance any longer, then the water rushing out of the waste will pull the water from the trap, and there won't be enough left to seal the trap. If you must have a longer run, you can buy a special 'deep-seal' trap.

Try to run the waste at a slope of about 1 in. in 3 foot or so (1:36); leave any steep angles for the last bit nearest the downpipe or gulley. You can bend the waste any amount on plan, although if you have too many bends eventually the pipe can clog.

Start by installing your sink or basin. There will be a large hole in the bottom, into which fits the waste. Wastes come in several sizes, and they come fitted with or without a hole for the overflow. The size of the waste can be checked with a tape measure, but sorting out the overflow is trickier.

Some appliances, such as wash basins, have a built-in

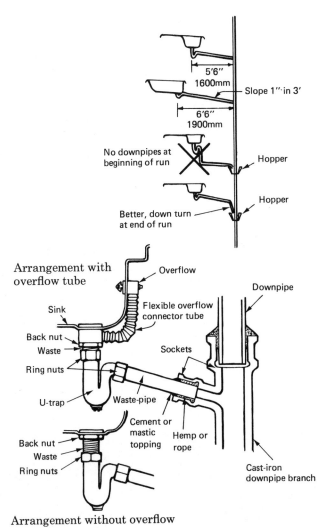

5'6"
1600mm

Slope 1" in 3'

6'6"
1900mm

No downpipes at
beginning of run

Hopper

Better, down turn
at end of run

Hopper

Arrangement with
overflow tube

Overflow

Downpipe

Sink

Flexible overflow
connector tube

Back nut

Waste

Sockets

Ring nuts

U-trap

Waste-pipe

Back nut

Cement or
mastic
topping

Hemp or
rope

Waste

Ring nuts

Cast-iron
downpipe branch

Arrangement without overflow

Fig. 57

117

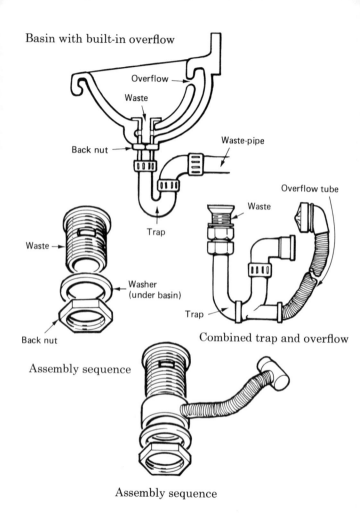

Basin with built-in overflow

Overflow

Waste

Back nut

Waste-pipe

Trap

Waste

Washer (under basin)

Back nut

Assembly sequence

Overflow tube

Waste

Trap

Combined trap and overflow

Assembly sequence

Fig. 58

overflow, as shown in the cross-section at the top of Fig. 58. These types need a waste fitting with an overflow slot ready cut.

Other appliances come with a slot cut into the back of the bowl, to which is fitted a plastic mouthpiece connected by a corrugated flexible pipe terminating in a ring which slips round the waste, as shown in the picture on the right of the above diagram. These also need a slotted waste.

Some baths have the same flexible pipe arrangement for the overflow, but the pipe terminates in a special trap; these appliances need a waste fitting without slot.

Lastly, some appliances don't have an overflow, and so the waste doesn't need a slot.

To fit the waste, take the waste in one hand, and unscrew the backing nut; put the washer and nut carefully aside, and fit the waste with a collar of putty. Knead the putty first, and then roll it into a $\frac{1}{2}$ in. (12 mm) snake which can fit round the waste, and be pushed to stick it down. Push the waste into the hole, and keeping it steady with one hand, push on first the overflow ring (where indicated) and then the washer and lastly the backing nut. An extra pair of hands, even the help of a child, can be helpful at this point, although it can be done by one person.

Screw the backing nut all the way till it gets tight. Now you will have to hold the waste steady while you tighten the back nut with a pair of adjustable pliers. Stick the open jaws of a pair of ordinary pliers into the grating at the top of the waste, and use a screwdriver for additional leverage as shown in Fig. 59. Tighten the back nut till it is firm, but don't overtighten it, or the porcelain or enamel may crack. Remove the surplus putty which will have squeezed out between bowl and waste.

Now we come to the trap. Traps come in different sizes to suit different waste diameters, and also to suit different waste-pipe diameters; check carefully before buying. Generally speaking, baths and sinks require a $1\frac{1}{2}$ in. (40 mm) diameter waste, basins can use a $1\frac{1}{4}$ in. (32 mm) waste.

As it is a new system, you will obviously be using plastic fittings. You can choose between a bottle trap or a U-bend

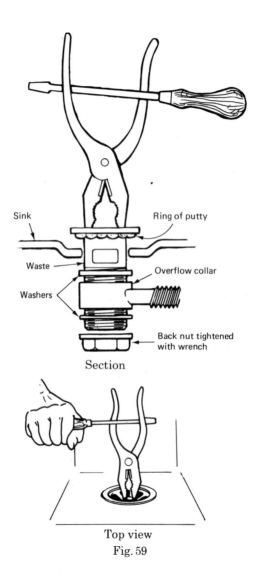

Sink

Ring of putty

Waste

Overflow collar

Washers

Back nut tightened
with wrench

Section

Top view
Fig. 59

trap (see Figs. 29 and 30). Both types disassemble for ease of cleaning if blocked; the choice of P- or S-types depends on where the waste-pipe goes after leaving the trap.

You must also choose between push fittings and glued fittings, as discussed in Section 6. Generally you cannot mix systems.

Whatever system you choose, buy trap and waste-pipe together. Start by screwing the trap on to the waste outlet; the ring nut of the trap is screwed on using hand-pressure only. Then cut the waste-pipe to length, using a tenon or junior hacksaw. Start by cutting it a little too long and fitting all the pieces together in a dry run (without using glue). It is easy to cut a little extra off if necessary, difficult to add it on if you have cut inaccurately.

Where the waste-pipe has to pass through a wall, use a cold chisel and hammer (see Section 5). For floors and holes through sink cabinets and the like, you will need a hole-cutter fitted on to the end of an electric drill. When forming a hole in the floor, take up the plank first, before drilling the hole, to check for any electric cables or gas/water pipes.

By now, the waste-pipe will have reached the point of discharge, which will be a gulley, a hopper or a downpipe.

Discharge into a hopper or gulley is easy; simply take the last bit of waste-pipe down to make sure the water doesn't spill over the sides.

Downpipes or soil-stacks are more of a problem. If the pipe is plastic, you will need a boss adaptor which sticks on to the plastic downpipe after you have cut a hole. Check the make (or better still the type of plastic) of the downpipe or soil-stack, which is usually shown on the connecting fittings. Each make of plastic requires its own special glue and fittings.

Cast-iron downpipes or soil-stacks cannot be adapted by the amateur. If the new waste-pipe is merely replacing an old one, you will have pulled the old one from a special branch fitting. The new pipe will terminate in the old socket, and you must caulk the space between the pipe and the socket with hemp or old bits of rope, hammered down with a blunt chisel and hammer. You finish off the caulking with some quickset cement or mastic.

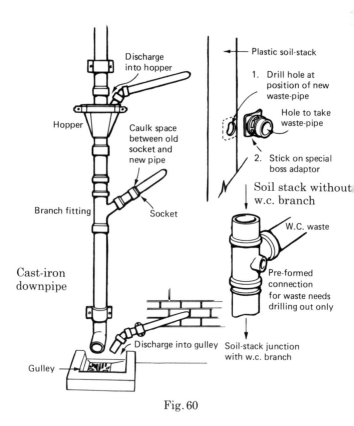

Discharge into hopper

Hopper

Caulk space between old socket and new pipe

Branch fitting

Socket

Cast-iron downpipe

Gulley

Discharge into gulley

Plastic soil-stack

1. Drill hole at position of new waste-pipe

Hole to take waste-pipe

2. Stick on special boss adaptor

Soil stack without w.c. branch

W.C. waste

Pre-formed connection for waste needs drilling out only

Soil-stack junction with w.c. branch

Fig. 60

If there is no handy gulley, hopper, plastic soil-stack or existing branch in a cast-iron downpipe, then I'm afraid you will have to call in a plumber to make the final connection.

20 Wash-basins

Materials

Obviously, you'll need a new wash basin. There are many different varieties and colours. If you're replacing an existing one, it saves tremendous amounts of time and money if you choose one with the holes for taps and waste in approximately the same place.

Again, by judiciously choosing a very similar model, you may save yourself having to buy new brackets. Otherwise, you must buy the brackets when buying the basin.

Decide whether you want entirely new taps (you can also buy a kit which puts a fancy plastic top on to your old tap) or to use the existing ones. Also decide on whether you need or want a new waste outlet, and possibly a new trap.

You will need a little putty for the waste, and possibly rawlplugs if you have to put in new brackets.

Tools

An adjustable wrench, a pair of adjustable pliers, a large screwdriver, possibly a basin wrench (this depends on whether you can take the basin down before removing the taps), a spirit level, and possibly a drill for making holes to take the plugs holding the brackets.

Time

Allow about half-an-hour for disconnecting the waste, about half-an-hour for disconnecting the taps at the swivel connectors; another half-hour for taking off each tap if you don't have to use a basin wrench, and three-quarters-of-an-hour if you do have to. Allow an hour if entirely new brackets have to be fitted, and a further hour for putting new taps and waste-pipe into the new basin (allow an extra hour or so for cleaning up an old waste and taps). Finally another hour for connecting the taps to the hot and cold water systems, and the waste to the waste-pipe. At worst the total time taken will be six hours or so, at best about three and a half hours.

The job

Using a pair of adjustable pliers, disconnect the waste-pipe from the trap (see Section 19 on waste-pipes). Use an adjustable wrench to disconnect the tail of the taps from the hot and cold water systems *after* you have turned the water off (see Sections 7 and 12). Now check to see how the bowl is held to the brackets and to the wall. Sometimes the bowl is screwed to the wall using long brass screws through holes in the back of the bowl and underneath. The bowl can also be held down by bolts to the brackets. Occasionally the bowl is fixed to its single bracket by the back nut of the waste. Undo all the relevant screws and bolts, and turn the bowl over.

With the adjustable wrench undo the back nut of the taps (see Section 12) and remove the taps. Similarly take off the back nut of the waste-pipe, and remove. Clean the hole (and the taps and waste-pipe if you are going to re-use these) of the remains of putty and old paint, using delicate blows of a hammer on an old screwdriver or cold chisel. Clear up the last traces using wire wool.

If you have to use new brackets, but can use the old waste-pipe and water supply pipes, assemble all these dry while the basin is resting on a chair (stacked with books to reach the right height). Mark where the basin is to sit on the wall. Take down the basin, and mark where the brackets are to sit, using the bowl as a template, and the spirit level in order to get them level. Drill and plug the brackets into place.

Insert the taps, with appropriate washers and a collar of putty (or a suitable rubber washer), into their holes, and screw the back nut into place (see Section 12 on taps). Place the basin on its brackets, and fix the bowl in place with the appropriate screws and bolts.

Insert the waste after fitting it out with a putty collar (see

Section 19 on wastes) and screw the back nut into place. Now connect the waste to the trap. If you have an existing metal U-trap, now is a good time to replace it with a plastic trap, which is so much easier to clean in case of blockage.

Connect the tap tails to the swivel joints at the end of the hot and cold water feeds, and tighten these.

That's all there's to it if you're replacing an existing basin. If the basin is installed into an entirely new position, then the connection to main drainage and water is more complicated. Turn to Sections 19 and 13 for further advice.

21 Putting in a sink

Materials

You'll obviously need a new sink. Sinks can be in a number of materials such as stainless steel, enamelled steel, fibreglass, porcelain.

Stainless steel is the most common material. It is usually somewhat thin and will drum under the impact of running water unless it is backed with a sheet of timber or with sprayed asbestos or fibreglass. It also shows up every dried-up water drop.

Enamelled steel can be finished in a variety of colours. The enamel will chip if anything is dropped on to the surface, such as a steak hammer or heavy knife.

Fibreglass is often used in caravans because of its light weight. In time it will scratch and become difficult to clean.

Porcelain sinks are very heavy, usually need a separate draining-board, and have no provision for taps.

Stainless steel, enamelled steel and fibreglass sinks come with either one or two bowls, left-hand, right-hand or double draining-boards, and in a variety of sizes as regards width and depth. If you're replacing an existing one, measure very carefully.

Think at the same time of renewing the taps and the waste system (see Sections 12 and 19), for which you will need the materials described in those sections.

If you're merely using existing taps and waste, you'll need a little putty.

Tools

An adjustable wrench, adjustable pliers, screwdriver, possibly a basin wrench, electrical pliers, drill with keyhole saws, hacksaw.

Time

Allow an hour and a half for turning off the water, and disconnecting the taps and waste-pipe. Another hour for figuring out how to get the old sink-top off, and half an hour to get the old taps off. Allow a half-hour for cleaning off putty and paint from the old waste and taps if you're re-using these. The next hour is for putting taps on to the new sink and fitting the new sink to the old stand. A final hour for connecting taps to cold water feed, and waste to trap. Altogether, at the worst about six hours, and about three if you're very lucky.

The job

Turn off the water supply to the taps. Disconnect the taps from the water feed pipes (see Section 12). Disconnect the waste from the trap (see Section 19). Now you must take the sink off its old base – this may require you to crawl underneath to spot all the screws and bolts holding it down. When the top is off, turn it over and take off the taps (see Section 12) and waste (see Section 19).

Now is a good time to install a stop-cock in each of the feed pipes to the taps.

If you are re-using old taps and waste, clear off putty and old paint using a stiff wire brush and possibly an old kitchen knife. When they're clean, fix them in the new sink (see Sections 12 and 19), and put the new sink on to the old base. Reconnect taps and waste to the old system.

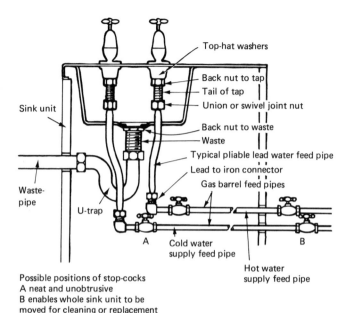

Top-hat washers

Back nut to tap

Tail of tap

Union or swivel joint nut

Back nut to waste

Waste

Typical pliable lead water feed pipe

Lead to iron connector

Gas barrel feed pipes

Sink unit

Waste-
pipe

U-trap

A   Cold water
supply feed pipe

B

Hot water
supply feed pipe

Possible positions of stop-cocks
A neat and unobtrusive
B enables whole sink unit to be
moved for cleaning or replacement

Typical sink unit seen from back

Fig. 61

When you're fitting a new sink in a new position, start by assembling or building the new base. Choose a location where the total waste run will not exceed 5 foot 6 inches (1600 mm) (see Section 19, on wastes). Run $\frac{1}{2}$ in. (15 mm) copper tube water feeds to the new sink, to terminate near the taps. If you can, use capillary fittings, and run the pipework in the thickness of the plaster. In order to do this, cut a groove in the plaster using your wide chisel. The groove should be about 1 in. (25 mm) wide and $\frac{3}{4}$ in. (19 mm) deep – you may have to cut the brickwork a little. If this seems too much like hard work, you can hire a machine to do this work. Bury the pipes, and cover with Polyfilla to achieve a smooth wall ready for re-painting.

Alternatively, run the pipework just above the skirting

board, and preferably behind kitchen worktop units. If you have to go through the sides of kitchen cabinets, use a keyhole saw attachment fitted to the end of an electric drill.

See Section 13 on the details of joining a cold or hot water feed to the taps.

The pipework for the waste can be run inside the kitchen cabinets, but bear in mind that waste-pipes take a fair amount of room, and that the space inside the cabinets will be that much less useful. For details on how to make up the waste, trap and waste-pipe, see Section 19.